SpringerBriefs in Economics

W0081970

More information about this series at http://www.springer.com/series/8876

Kostas Bithas · Panos Kalimeris

Revisiting the Energy-Development Link

Evidence from the 20th Century for Knowledge-based and Developing Economies

 Springer

Kostas Bithas
Panteion University
Athens
Greece

Panos Kalimeris
Panteion University
Athens
Greece

ISSN 2191-5504   ISSN 2191-5512   (electronic)
SpringerBriefs in Economics
ISBN 978-3-319-20731-5   ISBN 978-3-319-20732-2   (eBook)
DOI 10.1007/978-3-319-20732-2

Library of Congress Control Number: 2015957096

© The Author(s) 2016
This work is subject to copyright. All rights are reserved by the Publisher, whether the whole or part of the material is concerned, specifically the rights of translation, reprinting, reuse of illustrations, recitation, broadcasting, reproduction on microfilms or in any other physical way, and transmission or information storage and retrieval, electronic adaptation, computer software, or by similar or dissimilar methodology now known or hereafter developed.
The use of general descriptive names, registered names, trademarks, service marks, etc. in this publication does not imply, even in the absence of a specific statement, that such names are exempt from the relevant protective laws and regulations and therefore free for general use.
The publisher, the authors and the editors are safe to assume that the advice and information in this book are believed to be true and accurate at the date of publication. Neither the publisher nor the authors or the editors give a warranty, express or implied, with respect to the material contained herein or for any errors or omissions that may have been made.

Printed on acid-free paper

This Springer imprint is published by SpringerNature
The registered company is Springer International Publishing AG Switzerland

All things are exchanged for fire, and fire for all things, just as goods are for gold, and gold for goods.

—Heraclitus, Greek philosopher

Foreword

Our planet forms a complicated and multifaceted ecosystem. Some components are robust and stable, but others are vulnerable and fragile. Clearly, in an interdependent world, various forces are at work that may cause shifts and transformations of local or global ecosystems at different levels. Several megatrends may nowadays be observed in our century that form a threat to a sustainable development of our earth, such as depletion of resources, climate change, mass migration, large-scale urbanisation, socio-ethnic tensions, spatial-economic imbalances, health or disparities. One of the most pressing issues nowadays originates from environmental decay, caused by air, soil and water pollution and by climatological change. Several countries and regions—in both the development and the developing world—regard rightly environmental issues as one of the most challenging questions.

It goes without saying that environmental threats should not be seen in isolation, but in the broader context of globalisation forces, materials use and resource use, in particular energy resources. Clearly, energy is often the key to a better understanding and treatment of environmental deterioration. A balanced energy policy is usually directly and indirectly linked to environmental quality conditions. From that perspective, energy policy should not only be conceived of as a cost factor, but also as a source of new and beneficial opportunities in the environmental sector. Energy policy in a smart environment may have a double-dividend character and may be a critical facilitating factor for stimulating both economic efficiency and environmental effectiveness. To turn the 'bad' into the 'good' calls for innovative action strategies, from different and multidimensional—behavioural and policy—perspectives.

The present volume addresses many of the above-mentioned issues. It offers a new panorama for an integrated economic—environmental—energy policy and suggests unconventional roads to be travelled so as to ensure a sustainable future for our earth. It outlines a broad variety of policy, business and civic strategies and initiatives so as to cope with the various dilemma's involved with environmental threats. This study offers new insights that may lead to good practice guidelines for handling the complex energy—environmental—economic dilemma's. It is an eminent knowledge for students, researchers and practitioners.

Amsterdam   Peter Nijkamp
December 2015

Acknowledgments

The authors would like to thank Prof. Clive Richardson and Helga Stefansson, Panteion University, Athens, Greece, for their valuable assistance in editing and improving the present book. Furthermore, we would like to express our heartfelt gratitude to Prof. Jianguo Liu, Michigan State University, and Dr. Georgia Mavrommati, Dartmouth College, for their fruitful comments and suggestions during various stages of the present study. Parts of this book have been presented at the 11th International Conference of the European Society of Ecological Economics that was held in Leeds, United Kingdom, in 2015; and at the seventh Biennial conference of the US society for ecological economics that was held in the University of Vermont, Burlington, USA, in 2013. We thank the European Society of Ecological Economics for supporting and promoting this publication. We would also like to thank Anthony Doyle, Senior Editor of Springer UK, and Mr. Ravi Vengadachalam project coordinator for Springer Books for their encouragement and support throughout the period of writing this book until its publication.

Contents

Chapter 1
Introduction

There is good reason to doubt that past GDP growth per capita is entirely explained by capital accumulation or non-specific knowledge accumulation, as most growth theorists seem to believe [...] it is obvious that neither labor nor capital can function without inputs of energy.

Robert U. et al., (2013, pp. 80–81)

Twitter increased its turnover by 110 %[1] between 2013 and 2014. Facebook revenue for 2014 was $12.47 billion, marking an increase of 58 % in only a year.[2] In 2014, financial and insurance services represented 7.2 % ($1.26 trillion)[3] of the US GDP. It is commonly accepted that services will be producing more than 82 % of the US GDP, in the years ahead (Haksever and Render 2013). Twitter, Facebook, and, to a large extent, the entire service sector are economic activities with low energy and material requirements. Their substantial contribution to the US growth may support a vision of the decoupling of growth from energy and material use. However, before "consuming" Twitter, Facebook, and other "knowledge-based" services, the citizens of the USA need to "consume" food, housing, transport, and other "basic" products requiring substantial energy and material inputs. Notably, the relative prices of "basic" goods demonstrate a declining trend in relation to the prices of technologically advanced goods, like Twitter, Facebook, and other knowledge-based services. The trends in relative prices further enhance the vision of a decoupling of growth and natural resources, as modern economies shift towards the service sector. These decoupling estimates are all based on evaluations of the resource requirements at the level of aggregate GDP. As far as energy is concerned, the prevailing indicators, estimating the Energy-Economy link, follow the Energy/GDP prototype, reflecting the Energy Intensity (EI) of the economy. The prevailing EI indicators estimate the energy inputs consumed for the production of one unit of GDP and hence evaluate the efficiency of the economic system in its utilization of energy resources.

---

[1]Source: http://www.marketwatch.com/investing/stock/TWTR/financials (Accessed July 2015).
[2]Source: http://investor.fb.com/releasedetail.cfm?ReleaseID=893395 (Accessed August 2015).
[3]Source: http://selectusa.commerce.gov/industry-snapshots/financial-services-industry-united-states (Accessed August 2015).

© The Author(s) 2016
K. Bithas and P. Kalimeris, *Revisiting the Energy-Development Link*,
SpringerBriefs in Economics, DOI 10.1007/978-3-319-20732-2_1

Within this prevailing evaluation framework, the ultimate outcome of the economic system is envisaged to be represented by abstract monetary units aggregated in the GDP index. The link between the economy and resources is, then, evaluated as the efficiency of "transforming" resources into monetary units as reflected by the aggregate GDP.

Sane reasoning would suggest that the economic system produces goods whose consumption creates welfare/utility. Goods are consumed by human beings. The satisfaction of human needs is the ultimate objective of the economic process. Welfare/utility is a human perception arising from the use of goods; welfare is an individualistic phenomenon and therefore it can only be calculated at the level of individuals. Individuals are the superior entity of economic systems, a position broadly shared by all the schools of economic thought. The needs, the preferences and the utility of individuals are the fundamental variables of economic analysis. Within this analytical context, the assessment of economic welfare is a key issue in contemporary economics and serious analytical problems have arisen in this endeavor. The pressure from policy makers, in addition to operational necessities, has led to a broadly accepted compromise for the measurement of economic welfare/utility. "GDP per Capita" has been adopted as the appropriate proxy measure of economic welfare. "GDP per Capita" is the monetary index indicating the welfare arising from an economic system and enjoyed by human beings. The economic system creates economic welfare which is estimated as "GDP per Capita" and has been adopted by international organizations as the standard indicator for comparing economies. It depicts the ultimate outcome of the economy in monetary units, i.e. economic welfare. On the other hand, aggregate GDP can only offer a vision of the scale of the economy. Economies of the same aggregate GDP, as in the cases of the USA and China, may correspond to fundamentally different welfare levels. Indeed, the fundamentally different economic systems which produce different levels of welfare in the USA and China, are depicted in the substantially different levels of GDP per Capita i.e. 31,913 and 8,698 1990 GK$ in 2012 for the USA and China, respectively (The Conference Board Total Economy Database 2014). In fact, economies with the same GDP may reflect completely different economic systems with different economic structures. The essence of the economic system can only be approximated, in monetary terms, through the "GDP per Capita" index. "GDP per Capita" emerges as the appropriate monetary measure for the performance of the economic system. "GDP per Capita" gives a clear idea of the efficacy of economic processes in satisfying human needs.

Once "GDP per Capita" is adopted as the appropriate monetary-based indicator of the outcome of an economic system, then efficiency in utilizing natural resources can be estimated as *the resources required for the production of one unit of "GDP per Capita"*. The appropriate core indicator for the link between resources and the economy ought to be of the form **Natural Resources/[GDP per Capita]**, defined equivalently as **Natural Resources/Welfare**, or similarly as **Natural Resources/ Utility**, estimating the resource consumption required to create one unit of economic welfare/utility.

Within this volume we attempt to re-evaluate the link between the economy and energy resources. The ratio **Energy Resources/[GDP per Capita] = Energy Resources/Utility**, where *Utility = GDP per Capita*, is adopted as the core indicator of the proposed evaluation framework. An additional set of indicators is also estimated in order to delineate the profile of energy use in a number of national economies as well as for the global economy. Indeed, as the Energy-Economy link is intricate and multidimensional, we utilize a number of estimates in order to shed light on many important aspects of this complex interrelation.

Through the present volume we wish to contribute to an operational biophysical economic analysis. The economic system is envisaged as an integral part of a Coupled Human and Natural System (CHANS). The economic system works within the nexus of CHANS. It contributes to the objectives of the human system while it is constrained by natural laws and by the functions of the natural system (Liu et al. 2007).

The empirical evaluation of the Energy Intensity (EI) of the economy is feasible today due to recently developed datasets. Since 2000, recent scientific endeavors have established standardized and comparable datasets for the global economy as well as for a number of national economies. They have permitted an extensive analysis of the EI of several economies. These estimates form the empirical aftermath of an extensive and frequently hot analytical debate over the role of natural resources in the economic process. Prominent scholars including Nobel Laureate Robert Solow, Robert Ayres, Herman Daly and Nicholas Georgescu-Roegen have contributed to this theoretical debate with significant economic analyses. Hopefully, the recently established datasets permit, for the first time, a robust empirical investigation of the intricate link between the economy and natural resources. Although the empirical analysis is rather descriptive, useful results may now be drawn. Within this context we see the present volume as a modest contribution to the essential analysis of the link between energy resources and the economy.

The present book analyzes the long-run trends of Energy Intensity (EI) for the global economy (1900–2009), the USA (1870–2005), Japan (1878–2005), and 20 developed and developing economies with data ranging from 1965–2013. The estimates focus on the macro-trends which approximate the structural characteristics of the Energy-Economy link. Short run fluctuations of the EI, which reflect occasional causes, lie outside the focus of the current analysis. The structure of the book is the following:

Chapter 2 provides a brief history of energy use in human societies; Chap. 3 classifies the different forms of energy resources and presents the most prevailing energy aggregation methods; Chap. 4 provides a brief literature review on the Energy-Economy link and the decoupling effect; Chap. 5 analyzes the proposed methodological framework for the evaluation of the Energy-Economy link; Chap. 6 examines the empirical EI analysis at the global level and in 22 national economies; and finally, Chap. 7 provides the overall conclusions of the present volume. Annexes I-III provide additional information: the percentage changes of indicative variables and EI indicators (Annex I); an empirical estimate of the continental EI trends for both the standard and the proposed framework (Annex II); and finally (Annex III) the demographic trends, the GDP and GDP per capita growth trends, and Total DEC for the 22 countries examined as well as at the global aggregate level.

References

Ayres, R. U., Van den Bergh, J. C., Lindenberger, D., & Warr, B. (2013). The underestimated contribution of energy to economic growth. *Structural Change and Economic Dynamics, 27*, 79–88.

Haksever, C., & Render, B. (2013). *The important role services play in an economy*. Upper Saddle River: Financial Times (FT) Press. Available on-line at: http://www.ftpress.com/articles/article.aspx?p=2095734&seqNum=3. Accessed August 2015.

Liu, J. G., Dietz, T., Carpenter, S. R., Alberti, M., Folke, C., Moran, E., et al. (2007). Complexity of coupled human and natural systems. *Science, 317*, 1513–1516.

The Conference Board Total Economy Database. (2014). Available at: http://www.conference-board.org/data/economydatabase/.

Chapter 2
A Brief History of Energy Use in Human Societies

It is clear that there is some difference between ends: some ends are energeia, while others are products which are additional to the energeia.
Aristotle, Greek Philosopher, the first to describe the concept of energy.

Abstract Energy has always been among the most essential resources that endorses the progress, evolution and prosperity of human societies. This chapter aspires to provide a brief overview of historical evolution of energy use by human beings, from the discovery of fire and the agricultural revolution, to the industrial revolution and the domination of fossil fuels. By using historical evidence and brief diagrams, the narration provides a synoptic description of the unique and continuous quest of mankind for energy resources, unveiling the crucial role that energy still plays in modern economic systems as being the essential fuel of the economic process.

Keywords Energy history · Exosomatic metabolism · Organic economy · Fossil fuels economy · Industrial revolution

2.1 Introduction

Energy has always been among the most fundamental elements for the survival, reproduction and evolution of human society. The sun is the ultimate source of energy. Nonrenewable fossil fuels are formed by solar energy that has been captured over extremely long geological periods. What is more, renewable energy sources are directly (photovoltaic systems) and indirectly (wind, water, etc.) interrelated with the sun. Inevitably, almost all organisms rely, either directly or indirectly, on solar energy for their survival and maintenance. Life on earth would be impossible without the photosynthetic conversion of solar energy into plant biomass (Smil 1994). The sun provides approximately 1366 watts per square meter per second ($W/m^2/s$), hence, about 170,000 terawatts ($TW/m^2/s$) on the Earth's surface (Ruddiman 2001). In the food chain, solar energy flows are captured and converted through the complex process of photosynthesis. Part of this energy is

© The Author(s) 2016
K. Bithas and P. Kalimeris, *Revisiting the Energy-Development Link*,
SpringerBriefs in Economics, DOI 10.1007/978-3-319-20732-2_2

used by organisms, while a great proportion is lost as heat and a small portion is passed down the food chain as one organism digests another.[1] Apart from the food chain, intelligent human systems utilize the solar energy embodied in fossil fuels and the renewable energy sources as the essential power, the "engine" of modern civilization. The present chapter gives a brief overview of the historical evolution of energy utilization by the human societies.

2.2   The Mastery of Fire and Agriculture: The Organic Energy Economy

The very first milestone of mankind's utilization of energy was the mastery of fire. The utilization of fire for cooking and heating, using biomass (mainly wood) as fuel, dates back at least 4–500,000 years (Bowman et al. 2009). In addition, fire created light and thus improved safety in human settlements, a fact that promoted the expansion of habitation (Goudsblom 1992; Fouquet 2011). The burning of wood and other forms of biomass eventually led to the discovery of ovens which, besides cooking, permitted the early forms of crafting. Ovens made it possible to produce pottery and to refine metals from ore.[2] Early humans lived a largely nomadic existence, closely in synchrony with the change of seasons and periodic plant growth.

The next milestone of mankind was the Agricultural Revolution (Heinberg 2011). The introduction of agriculture increased the amount of available food, permitting the first permanent human settlements, which caused a substantial increase of human population. Water and wind power were the next essential steps in the evolution of the human conquest of energy. The watermill was invented about 2500 years ago (Lucas 2006). Using both the water and the windmills, humans managed to master the water and air power necessary to meet their needs for crushing grain (wheat, etc.) in order to produce flour, crushing olives for olive oil production, tanning leather, smelting iron, sawing wood, and so on (Reynolds 1983). However, despite the improvements in energy use and the exploitation of several energy resources, the rapid growth of population in Europe about a thousand years ago—as a result of this progress—led to dramatic pressures on land for cultivation, and forests were being encroached upon to provide more land (Georgescu-Roegen 1984; Fouquet 2011).

This first era of mankind's quest for new energy resources, from the early discovery of fire to the agricultural (and farming) revolution, could be briefly described as the **Organic Energy Economy** (Fouquet 2011). This solar-based energy economy was intimately based on intensive land use and biomass

---

[1]Energy Literacy. Essential principles and fundamental concepts for energy education. U.S. Department of Energy. Available on-line at: http://www1.eere.energy.gov/education/pdfs/energy_literacy_2.0_low_res.pdf. (Accessed March 2015).
[2]Ibid.

consumption. This pre-industrial economy, dominated by the so-called "*somatic energy regime*" (McNeill 2000), was an era in which "*endosomatic metabolism*" and biomass consumption were the predominant elements of the "*agrarian metabolic regime*" (Krausmann 2011). Inevitably, the organic energy economy was limited to the consumption of energy at the rate that solar energy can be converted into useful goods and services. In this context, population growth and the limited land availability imposed crucial restrictions upon further economic growth and gradually forced a transition towards a new energy regime, the era of fossil fuels (Fouquet 2011; Krausmann 2011).

2.3 Transition to the Fossil Fuel Economy

The milestone that determined the transition from the organic economy to the fossil fuel economy, the invention that characterized the era called "The Industrial Revolution", was the steam engine. The unique process that the steam engine initiated was the conversion of chemical energy (heat) into mechanical energy (motion) (McNeill 2000). The biomass energy stocks accumulated in the earth's crust for hundreds of millions of years were now available to serve human needs for the first time in mankind's history, to such an extent that the dawn of the fossil fuel era was about to begin. While the early steam engine was mainly used for pumping water out of coal mines, it soon became—thanks to the efficiency improvements made by James Watt, a Scottish inventor and mechanical engineer—a valuable tool which increased human muscle and animal power for extracting more coal, drove the manufacturing industry, moved ships and trains, and laid the foundation for today's complex and energy intensive human (economic) systems (Fouquet 2011).

During the 18th century, many industries had already substituted wood-fuels with coal, while heating services made the transition from organic biomass to fossil fuels by the beginning of the 19th century.[3] Specifically, between 1650 and 1740, the real prices of wood-fuel increased substantially, which encouraged its progressive substitution with coal (Fouquet 2008, 2011). The timing of this substitution was absolutely essential, given the fact that during the second half of the 17th century the harvesting of forest trees had to be regulated, even restricted, in England and elsewhere in Europe (Georgescu-Roegen 1984). Wrigley (1988) suggests that, by 1800, had the British economy been dependent on wood-fuel, a surface area equivalent to the whole of Britain would have had to be coppiced every year in order to supply the energy demands of the economy. On the other hand, wind and water power provided only one-tenth of the total power of the British economy in 1800 (Fouquet 2006). By 1900, steam engines provided two-thirds of all power services; the expansion of the railway network provided more than 90 % of goods

---

[3]Three quarters of the energy requirements of British economy were used for heating services (households, buildings, industry) (Fouquet 2008).

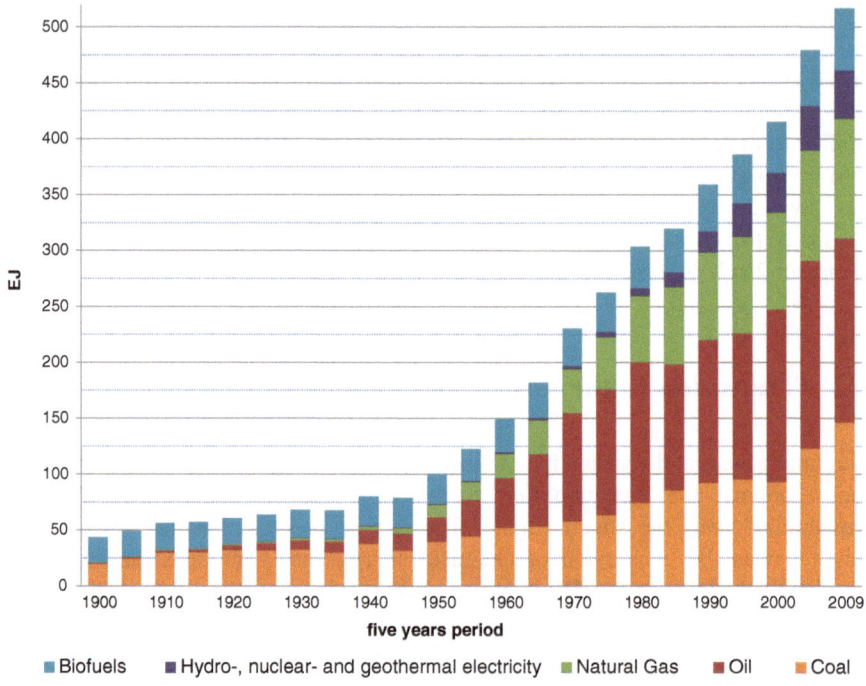

Fig. 2.1 The world's total primary energy supply for 1900–2009

transportation on land, while steam ships were carrying about 80 % of all freight cargos at sea (ibid).

Remarkably, the growing demand for coal in the 19th century raised concerns about coal scarcity and its consequences for the production process (Jevons 1865). However, new technological improvements managed to maintain a constant coal supply, and simultaneously kept prices low (Fouquet 2006). Furthermore, the introduction of new energy resources, such as petroleum and petroleum by-products, enhanced the fossil energy mix. The major invention that really promoted the use of refined oil was the internal combustion engine. While the process of refining crude oil[4] provided the foundation for the oil age, it was following the invention and development of internal combustion engines in Germany, after 1880, that the use of oil took off (McNeill 2000). Peak oil production was reached in the US in late 1960s, and the heightened concerns about security in maintaining a constant energy supply induced by the oil shocks of 1973 and 1979, led to a rapid increase in natural gas use. Clearly, after the 1970s natural gas consumption

---

[4]Discovered by James Young in the 1850s, while Edwin Drake, in 1859, managed to successfully drill for oil through deep rock.

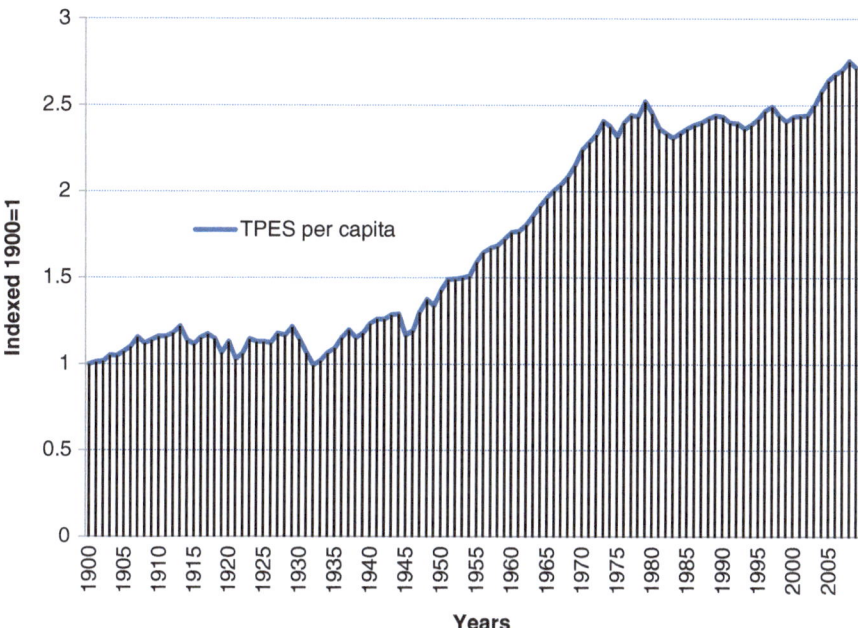

Fig. 2.2 The world's per captia total primary energy supply 1900–2009

increases dramatically (Fig. 2.1). Coal consumption overcomes biofuel consumption from the very first years of the 20th century, while oil consumption takes the lion's share from coal in the early 1960s. The use of natural gas increases dramatically after WWII, while hydro, nuclear and geothermal electricity use increases constantly from the early 1970s. Finally, and remarkably so, bio-fuel consumption steadily increases throughout 1900–2000, with a further acceleration of incremental trends occurring in the early 2000s (Figs. 2.1 and 6.1a).

Figure 2.2 displays the trajectory of the world's per capita Total Primary Energy Supply (TPES), for 1900–2009. An extraordinary and unparalleled rise in per capita energy use occurred after WWII. This trend shows signs of stabilization from 1980–2000; however, from 2000–2009, a further acceleration is seen again.

References

Bowman, D. M., Balch, J. K., Artaxo, P., et al. (2009). Fire in the Earth system. *Science, 324* (5926), 481–484.

Fouquet, R. (2008). *Heat, power and light: Revolutions in energy services.* Cheltenham, UK: Edward Elgar Publishing.

Fouquet, R. (2011). A brief history of energy. In J. Evans & L. C. Hunt (Eds.), *International handbook on the economics of energy* (pp. 1–19). Cheltenham, UK:Edward Elgar Publishing.

Fouquet, R., & Pearson, P. J. (2006). Seven centuries of energy services: The price and use of light in the United Kingdom (1300–2000). *The Energy Journal, 27*(1), 139–177.

Georgescu-Roegen, N. (1984). Feasible recipes versus viable technologies. *Atlantic Economic Journal, 12*(1), 21–31.

Goudsblom, J. (1992). The civilizing process and the domestication of fire. *Journal of World History, 3*(1), 1–12.

Heinberg, R. (2011). *The end of growth: Adapting to our new economic reality.* Gabriola Island, BC: New Society Publishers.

Jevons, W. S. (1865). *The coal question: an inquiry concerning the progress of the nation, and the probable exhaustion of our coal-mines.* Macmillan.

Krausmann, F. (2011). *The socio-metabolic transition. Long term historical trends and patterns in global material and energy use.* Social Ecology Working Paper 131. Available at: http://www. uni-klu.ac.at/socec/downloads/WP131FK_webversion.pdf.

Lucas, A. (2006). *Wind, Water, Work. Ancient and Medieval Milling Technology.* The Netherlands: Brill Academic Publishers.

McNeill, J. R. (2000). *Something new under the sun: An environmental history of the twentieth-century world* (the global century series). New York: WW Norton and Company.

Reynolds, L. G. (1983). The spread of economic growth to the third world: 1850–1980. *Journal of Economic Literature, 21*(3), 941–980.

Ruddiman, W. F. (2001). *Earth's climate: Past and future.* London: Freeman.

Smil, V. (1994). *Energy in world history.* Boulder, CO: Westview Press.

Wrigley, E. A. (1988). The limits to growth: Malthus and the classical economists. *Population and Development Review, 14*, 30–48.

Chapter 3
Defining the Energy Resources

Hitherto it is questionable if all the mechanical inventions yet made have lightened the day's toil of any human being. They have enabled a greater population to live the same life of drudgery and imprisonment, and an increased number of manufacturers and others to make fortunes.

John Steward Mill, 1848, pp. 756–757

Abstract What is energy? This chapter endeavors to answer the anterior question by providing definitions of energy from the viewpoint of physics, and discusses the various forms that energy takes so as to serve the human needs. The issue of energy aggregation is of paramount importance for comparing and evaluating different forms of energy. In this context, the present chapter describes and classifies the most widely utilized methodologies in energy measurement.

Keywords Energy forms · Energy measurement · Energy aggregation · Primary energy resources

3.1 What Is Energy, Really?

The word "energy" is derived from the Greek word "energeia", which combines two roots meaning "at" and "work", a metaphysical concept employed by Aristotle to define the "action towards a goal" (Cleveland and Morris 2009—p. 166).

Seen from the viewpoint of physics, energy is a fundamental physical concept that can be described as the potential ability of a system to influence changes in other systems. Two forms of change can be induced: (Cleveland and Morris 2009):

– work (forced directional displacement)
– heat (chaotic displacement/motion of system microstructure)

Before the Industrial Revolution there were mainly four sources of mechanical work of economic significance (Ayres and Warr 2009—p. 90):

– Human labor (muscle work)
– Animal labor (muscle work)
– Water power (water mills)
– Wind power (wind mills)

© The Author(s) 2016
K. Bithas and P. Kalimeris, *Revisiting the Energy-Development Link*,
SpringerBriefs in Economics, DOI 10.1007/978-3-319-20732-2_3

Today, mechanical work is provided mainly by prime movers[1] which can be classified in two broad categories (*ibid*): turbines (hydraulic and steam); and internal combustion engines (spark ignition gasoline engines; compression ignition diesel engines; and gas turbines).

This capacity to do either heat or mechanical work is used to perform useful functions, such as heating and cooling, mechanical motion and transportation, driving machinery, lighting, food production, infrastructure building, and so on. The following section aims to cast light on the different forms of energy available and to classify them into distinct categories.

3.2   Forms of Energy

Energy exists in many different forms, which could be briefly represented in six broad categories, in terms of physical science and especially mechanical theory (Cleveland and Morris 2009).

3.2.1   Mechanical (Potential and Kinetic[2])

Mechanical energy can be broadly classified into potential and kinetic energy. Classic examples of potential mechanical energy are the gravitational energy embedded in water falling from a higher to a lower point (e.g. hydroelectricity production from a dam), as well as other forms such as electrostatic and magnetic fields. Mechanical energy can be also converted into kinetic energy (the work required to accelerate an object to a given speed), with the internal combustion engine as a representative example of this transformation.

3.2.2   Thermal

Thermal energy is the kinetic energy associated with the motion of atoms and molecules within a substance (i.e. heat). Besides the functions of heating and cooking, thermal energy is broadly used for producing electricity (as in nuclear and coal power plants).

---

[1]An engine or a device by which a natural energy resource (an energy vector) is converted into mechanical power (Cleveland and Morris 2009—p. 403).

[2]In physics, mechanical energy is the sum of the kinetic energy and potential energy of an object (Cleveland and Morris 2009—p. 320).

3.2.3 Electrical (and Electromagnetic)

Electricity is a fundamental form of energy in today's complex human systems. It consists of electrons and protons with opposite charges. Electrical energy produces many other energy forms, such as thermal (heat, light), kinetic, and magnetic, as well as chemical (i.e. electrolysis).

3.2.4 Radiant

Radiation is energy in the form of electromagnetic waves. Solar energy is a representative example of radiant energy.

3.2.5 Chemical

The energy produced or absorbed during the process of a chemical reaction. Some typical examples of chemical energy production are food digestion by living organisms which provide the necessary energy for their survival, fire, batteries (as stored chemical energy that provides electricity), fuel-cells, and so on.

3.2.6 Atomic (Nuclear)

Energy released by radioactive decay, through a nuclear reaction or in the course of fission (splitting) or fusion (fusing) of atomic nuclei. Nuclear energy is used mainly for electricity production, while it can produce electromagnetic and kinetic energy, as well.

3.3 Primary and Secondary Energy Forms

An important classification of energy is the distinction between primary and secondary energy forms:

Primary Energy forms: primary energy is the energy embodied in natural resources prior to undergoing any human-induced conversions or transformations (Cleveland and Morris 2009—p. 402). This category includes all forms of energy found in nature without any previous conversion or transformation process. From a thermodynamic perspective, primary energy resources cannot be produced by human systems. The following natural assets and functions are considered primary energy sources: fossil fuels (crude oil, coal, and natural gas), natural uranium, the gravitational energy of water, tidal energy, biomass and geothermal energy, and flows such as solar and wind energy.

Secondary Energy forms: With the term "secondary energy forms", we define all those forms of energy that can be directly used for serving human purposes and transformed into useful forms of energy, through conversions of primary energy sources. Electrical energy is the most indicative secondary form of energy. Other common forms of secondary energy are produced through fuel refining (e.g. gasoline from crude oil), synthetic fuels (hydrogen fuel) and any transformation of fossil fuels into electrical, kinetic, or thermal energy (Cleveland and Morris 2009—p. 451).

3.4   Energy Measurement and Aggregation

An important issue, which arises from the need to "homogenize" the varied and "heterogeneous" fuel types and energy categories, is the method of energy measurement (Cleveland et al. 2000; Stern 2010). The appropriate selection of a measuring unit is of crucial importance for the conversion of different energy types into a comparable unit which could allow the aggregation and comparison of different energy resources (Stern 2010). This section attempts to give a brief representation of the most common thermal measuring units and methods utilized for energy aggregation in the vast majority of international reports and research studies.

The simplest method of energy aggregation, and the most common among relevant studies, is to add up the different energy forms according to their thermal equivalents (Cleveland et al. 2000). The most commonly used methods of thermal equivalent energy aggregation are the following:

British thermal unit (Btu)

The British thermal unit is defined as the average amount of energy required to induce a change in temperature of 1 °F in one pound of pure liquid water, and is equivalent to about 1055 joules, or 252 calories (Cleveland and Morris 2009—p. 62–63).

Tons (or tons) of Oil equivalents (Toe)

Tons of oil equivalents is a measure of energy used to relate different fuels to the equivalent amount of energy released by burning one ton of crude oil, approximately 2 Giga-Joule (GJ), (Cleveland and Morris, 2009—p. 522). Many studies use the million tons of oil equivalents which is abbreviated as *mtoe*.

Joule (J)

The joule is the basic unit of energy in the meter-kilogram-second system and is translated as the amount of work done by a force of one Newton acting through a distance of one meter in the direction of the force (Cleveland and Morris 2009—p. 279).

Watt-hour

Watt per hour (Wh) is the unit of energy which is equal to the work done by one watt over a time of one hour. The kilowatt per hour (kWh), equivalent to 3.6 × 106 J (3600 kJ or 3.6 MJ), is the prevailing measuring unit for electricity. Yearly electricity usage is often given in units of kilowatt-hours per year (kWh/year).

In the present volume, energy is mainly measured and aggregated in Exa-joule (EJ) and Peta-joule (PJ) for the long run datasets of the global economy, the USA, and Japan (Sects. 6.2–6.4), and in million tons of oil equivalents (mtoe, in the short run estimates of the USA and Japan (Sect. 6.5), as well as in the case of the 10 developed and the 10 developing countries studied (Sects. 6.6–6.7).

References

Ayres, R. U., & Warr, B. (2009). *The economic growth engine: how energy and work drive material prosperity*. London: Edward Elgar Publishing.

Cleveland, C. J., Kaufmann, R. K., & Stern, D. I. (2000). Aggregation and the role of energy in the economy. *Ecological Economics, 32*(2), 301–317.

Cleveland, C. J., & Morris, C. (2009). *Dictionary of energy*. London: Elsevier.

Stern, D. (2010). Energy quality. *Ecological Economics, 69*, 1471–1478.

Chapter 4
Exploring the Energy-Economy Link

> *I'd put my money on the sun and solar energy. What a source of power! I hope we don't have to wait 'til oil and coal run out before we tackle that.*
>
> Thomas Edison (1847–1931)

Abstract Energy emerges as an indispensable input in the production process and the economy depends on energy resources. The evaluation of the intensity of this dependency is a crucial analytical problem which attracted the interest of scientists early on. Recently developed datasets permitted the empirical evaluation of the link between energy and the economy. The Material Framework Analysis (MFA) has led to a number of empirical studies concerning several economies for long periods. The key indicator in these empirical studies is the "energy resources consumed for producing one unit of GDP". Evidently the rationale behind this indicator envisages the economy as an engine producing GDP reflected in the aggregate value of all goods produced within the economic system. On the basis of this key indicator a large number of recent studies identify a decoupling between energy and growth.

Keywords Material flow analysis · Energy intensity · Decoupling effect

4.1 A Literature Review

Are our economies becoming less dependent on energy resources? (Weizsäcker et al. 1997; Herring 2006) Is current growth undertaking a transition towards declining Energy Intensity (EI)? (Cornillie and Fankhauser 2004; Schäfer 2005) Will the transition to a modern service economy delink the production process from energy resources use and lead to environmental improvement? (Powell and Snellman 2004; Kander 2005; Wölfl 2005; Spohrer and Maglio 2008) Has the long-standing dialogue on the constraints of economic growth imposed by the scarcity of natural resources (including energy) been resolved in favor of the "optimistic school"? (Solow 1974; Bithas and Nijkamp 2006; Bithas 2008, 2011)

There are many ways to examine the above questions. One possibility is to use real data in order to estimate contemporary trends in the relationship between

© The Author(s) 2016
K. Bithas and P. Kalimeris, *Revisiting the Energy-Development Link*,
SpringerBriefs in Economics, DOI 10.1007/978-3-319-20732-2_4

economic growth and energy inputs. In this context, the Economy-Wide Material Flow Analysis[1] (MFA) has emerged as a framework of substantial maturity and methodological accuracy, showcasing a wide range of publicly accessible databases available for empirical analysis and for comparisons between methods and results (Fischer-Kowalski et al. 2011). More specifically, three major types of studies can be identified in the relevant literature concerning the Energy-Economy link (Madlener 2011):

- Studies employing sophisticated econometric techniques.
- Studies employing elasticities for computing the responsiveness of energy consumption to GDP growth.
- Studies estimating the energy flows required for the production of one unit of GDP. These requirements are defined as the Energy Intensity of an economy estimated through the E_t/GDP_t ratio, where E_t indicates the annual energy inputs.

The present volume explicitly deals with the third category of studies which estimate the energy input required for the production of one unit of GDP, a broadly accepted indicator within the MFA framework. In essence, the Energy Intensity indicator provides the empirical estimation of the so-called decoupling of the economy from energy use. There exist many alternative applications of the E_t/GDP_t prototype that could be briefly summarized in four categories:

- Total Energy Consumption $(TEC)_t/GDP_t$ (Cleveland et al. 1984; Kauffman 1992, 2004)
- Total Primary Energy Supply $(TPES)_t/GDP_t$ (Krausmann et al. 2009)
- Domestic Energy Consumption $(DEC)_t/GDP_t$ (Haberl et al. 2006)
- (Useful Work)$_t/GDP_t$ (Ayres and Warr 2009; Serrenho et al. 2014)

As one of the most widely cited macroeconomic indicators for measuring sustainability through estimates of the decoupling effect, the E/GDP ratio has been the focus of a significant number of published studies. The majority of these studies examine the decoupling effect at the level of a single country (Bullard and Foster 1976; Ostblom 1982; Bossanyi 1979; Kaufmann 1992; Garbaccio and Jorgenson 1999; Wing 2008) or of a group of countries (Reister 1987; Howarth et al. 1993; Mulder and de Groot 2004; Markandya et al. 2006; Warr et al. 2010), while a few have recently attempted an investigation of the global Energy-Economy link (Krausmann et al. 2009; UNEP 2011; Bithas and Kalimeris 2013). Concerning studies which estimate decoupling for a single country, the vast majority maintain that energy consumption grew at a much slower pace despite the significant increase in GDP (Östblom 1982, 1993; Garbaccio and Jorgenson 1999; Stern 2011). Similarly, empirical studies using groups of developed countries show a

---

[1]Or Accounting.

clear decline in the Energy/GDP ratio over most of the last half of the 20th century (Nilsson 1993; Dincer 1997). According to MacKillop (1990), post-war energy intensity (at least for OECD countries) can be separated into two periods: before 1973 when energy and economic growth were coupled; and after 1973 when they were clearly decoupled. Following similar reasoning, various reports have identified a permanent decoupling between energy and economic growth (IEA 1982; World Bank 1992; OECD 2002). More recent empirical studies, estimating decoupling at the global level, further support a delinking of GDP from the use of energy; a trend that continues unchanged until today (Krausmann et al. 2009; UNEP 2011). As a consequence, it seems that modern economic growth has entered a period of transition towards less dependence on energy resources (Ross et al. 1987; Ziolkowska and Ziolkowski 2011).

Taking a different perspective, certain studies (de Bruyn and Opschoor 1997) claim that this "delinking" trend does not prove to be persistent, an argument that nourishes scepticism over the decoupling effect (Auty 1985; Cleveland and Ruth 1999; Herring 2006; Kander 2005). Similarly, an early study on decoupling (Bullard and Foster 1976) argues that substantial technological and lifestyle changes may be required if energy and economic growth are to be decoupled, while emphasizing the importance of population and GDP per Capita growth rates in the decoupling debate.

4.1.1   Recent Criticisms of the Prevailing EI Framework

Contemporary analysis is mainly directed towards criticizing the methods and the techniques related to the appropriate energy measurement (Ayres et al. 2003; Ayres and Warr 2009; Warr and Ayres 2010; Serrenho et al. 2014). In this direction, many deal with the proper energy aggregation (Cleveland et al. 2000; Stern 2010, 2011), while others analyse the substitution trends among different energy resources with qualitative differences (Kaufmann 1992, 2004). All these efforts deal mainly with the appropriateness of the nominator of the E_t/GDP_t prototype, while the relevant literature completely ignores the important implications and constraints raised by the prevailing use of GDP as the denominator in the vast majority of the published studies. The GDP index has been severely and extensively criticized by many distinguished scholars with regard to its inability to reflect the actual welfare that the economic system creates (Ayres et al. 1996; van den Bergh 2010; Daly 2013; Costanza et al. 2014). In this context, there are recent empirical studies estimating the Energy-Economic Welfare link that support a coupling relationship between energy flows and the global economy (Bithas and Kalimeris 2013).

References

Auty, R. (1985). Materials intensity of GDP: Research issues on the measurement and explanation of change. *Resources Policy, 11*(4), 275–283.

Ayres, R. U., & Warr, B. (2009). *The economic growth engine: How energy and work drive material prosperity*. UK: Edward Elgar Publishing.

Ayres, R. U., Ayres, L. W., & Martinas, K. (1996). *Eco-thermodynamics: Exergy and life cycle analysis*. INSEAD, Center for the Management of Environmental Resources. Working Paper 961041.

Ayres, R. U., Ayres, L. W., & Warr, B. (2003). Exergy, power and work in the US economy, 1900–1998. *Energy, 28*(3), 219–273.

Bithas, K. (2008). Tracing operational conditions for the ecologically sustainable economic development: The Pareto optimality and the preservation of the biological crucial levels. *Environment, Development and Sustainability, 10*(3), 373–390.

Bithas, K. (2011). Sustainability and externalities: Is the internalization of externalities a sufficient condition for sustainability? *Ecological Economics, 70*(10), 1703–1706.

Bithas, K., & Kalimeris, P. (2013). Re-estimating the decoupling effect: Is there an actual transition towards a less energy-intensive economy? *Energy, 51*, 78–84.

Bithas, K., & Nijkamp, P. (2006). Operationalising ecologically sustainable development at the microlevel: Pareto optimality and the preservation of biologically crucial levels. *International Journal of Environment and Sustainable Development, 5*(2), 126–146.

Bossanyi, E. (1979). UK primary energy consumption and the changing structure of final demand. *Energy Policy, 7*(3), 253–258.

Bullard, C. W., & Foster, C. Z. (1976). On decoupling energy and GDP growth. *Energy, 1*, 291–300.

Cleveland, C. J., & Ruth, M. (1999). Indicators of dematerialization and the materials intensity of use. *Journal of Industrial Ecology, 2*(3), 15–50.

Cleveland, C. J., Costanza, R., & Hall, C. A. S., Kaufmann, R. (1984). Energy and the US economy: A biophysical perspective. *Science, 255*, 890–897.

Cleveland, C. J., Kaufmann, R. K., & Stern, D. I. (2000). Aggregation and the role of energy in the economy. *Ecological Economics, 32*(2), 301–317.

Cornillie, J., & Fankhauser, S. (2004). The energy intensity of transition countries. *Energy Economics, 26*, 283–295.

Costanza, R., et al. (2014). Time to leave GDP behind. *Nature, 505*, 283–285.

Daly, H. E. (2013). A further critique of growth economics. *Ecological Economics, 88*, 20–24.

de Bruyn, S. M., & Opschoor, J. B. (1997). Developments in the throughput-income relationship: Theoretical and empirical observations. *Ecological Economics, 20*(3), 255–268.

Dincer, I. (1997). Energy and GDP analysis of OECD countries. *Energy Conversion and Management, 38*(7), 685–696.

Fischer-Kowalski, M., Krausmann, F., Giljum, S., Lutter, S., Mayer, A., Bringezu, S., et al. (2011). Methodology and indicators of economy-wide material flow accounting. State of the art and reliability across sources. *Journal of Industrial Ecology, 15*(6), 855–875.

Garbaccio F. R., Ho S. M., & Jorgenson W. D (1999). Why has the energy-output ratio fallen in China? Cambridge: Kennedy School of Government, Harvard University. http://www.hks.harvard.edu/m-rcbg/ptep/energy-ratio.pdf. Accessed in 2012.

Haberl, H., Weisz, H., Amann, C., Bondeau, A., Eisenmenger, N., Erb, K. H., et al. (2006). The energetic metabolism of the European Union and the United States: Decadal energy input time-series with an emphasis on biomass. *Journal of Industrial Ecology, 10*(4), 151–171.

Herring, H. (2006). Energy efficiency—A critical view. *Energy, 31*, 10–20.

Howarth, B. R., Schipper, L., & Adersson, B. (1993). The structure and intensity of energy use: Trends in 5 OECD nations. *Energy Journal, 14*(2), 27–45.

IEA. (1982). *The world outlook for energy to 2020*. Paris: IEA.

Kander, A. (2005). Baumol's disease and dematerialization of the economy. *Ecological Economics, 55*(1), 119–130.

Kaufmann, R. K. (1992). A biophysical analysis of the energy/real GDP ratio: Implications for substitution and technical change. *Ecological Economics, 6*(1), 35–56.

Kaufmann, R. K. (2004). The mechanisms for autonomous energy efficiency increases: A cointegration analysis of the US energy/GDP ratio. *The Energy Journal, 25*, 63–86.

Krausmann, F., Gingrich, S., Eisenmenger, N., Erb, K. H., Haberl, H., & Fischer-Kowalski, M. (2009). Growth in global materials use, GDP and population during the 20th century. *Ecological Economics, 68*(10), 2696–2705.

MacKillop, A. (1990). On decoupling. *International Journal of Energy Research, 14*, 83–105.

Madlener, R. (2011). The economics of energy in developing countries. In L. C. Hunt (Ed.), *Evans J* (pp. 740–758). Edward Elgar: International handbook on the economics of energy.

Markandya, A., Pedroso-Galinato, S., & Streimikiene, D. (2006). Energy intensity in transition economies: Is there convergence towards the EU average? *Energy Economics, 28*, 121–145.

Mulder, P., & de Groot, H. L. F. (2004). Decoupling economic growth and energy use: An empirical cross-country analysis for 10 manufacturing sectors. Tinbergen Institute. Discussion Paper No. 04-005/3. Available at SSRN: http://ideas.repec.org/p/dgr/uvatin/20040005.html. Accessed in December 2012.

Nilsson, J. L. (1993). Energy intensity trends in 31 industrial and developing countries 1950–1988. *Energy, 18*(4), 309–322.

Östblom, G. (1982). Energy use and structural changes: Factors behind the fall in Sweden's energy output ratio. *Energy Economics, 4*(1), 21–28.

Östblom, G. (1993). Increasing foreign supply of intermediates and less reliance on domestic resources: The production structure of the Swedish economy, 1957–1980. *Empirical Economics, 18*(3), 481–500.

Reister, D. B. (1987). The link between energy and GDP in developing countries. *Energy, 12*(6), 427–433.

Schäfer, A. (2005). Structural change in energy use. *Energy Policy, 33*, 429–437.

Serrenho, A. C., Sousa, T., Warr, B., Ayres, R. U., & Domingos, T. (2014). Decomposition of useful work intensity: The EU (European Union)-15 countries from 1960 to 2009. *Energy, 76*, 704–715.

Solow, R. M. (1974). The economics of resources or the resources of economics. *The American Economic Review, 64*(2), 1–14.

Spohrer, J., & Maglio, P. P. (2008). The emergence of service science: Toward systematic service innovations to accelerate co-creation of value. *Production and Operations Management, 17*, 238–246.

Stern, D. (2011). The role of energy in economic growth. *Annals of the New York Academy of Sciences, 1219*, 26–51.

UNEP. (2011). Decoupling natural resource use and environmental impacts from economic growth. In M. Fischer-Kowalski, M. Swilling, E. U. von Weizsäcker, Y. Ren, Y. Moriguchi, W. Crane, et al. (Eds.), A Report of the Working Group on Decoupling to the International Resource Panel, United Nations.

Warr, B., Ayres, R., Eisenmenger, N., Krausmann, F., & Schandl, H. (2010). Energy use and economic development: A comparative analysis of useful work supply in Austria, Japan, the United Kingdom and the US during 100 years of economic growth. *Ecological Economics, 69*(10), 1904–1917.

Weizsäcker, E., von Lovins, A., & Lovins, L. H. (1997). *Doubling wealth—Halving resource use.* London: Earthscan.

Wing, I. S. (2008). Explaining the declining energy intensity of the U.S. economy. *Resource and Energy Economics, 30*(1), 21–49.

Wölfl, A. (2005). *The Service Economy in OECD countries.* STI Working Paper. Statistical Analysis of Science, Technology and Industry JT00178454, OECD. http://www.oecd.org/officialdocuments/publicdisplaydocumentpdf/?cote=DSTI/DOC%282005%293&docLanguage=En. Accessed in 2012.

World Bank. (1992). World development report. Development and the environment. New York: Oxford University Press. Available on-line in: http://wdronline.worldbank.org/worldbank/a/c. html/world_development_report_1992/chapter_6_energy_industry. Accessed in May 2012.

OECD. (2002). Indicators to measure decoupling of environmental pressure from economic growth. SG/SD (2002)1/FINAL, Paris: OECD. Available on-line at: http://www.oecd.org/officialdocuments/publicdisplaydocumentpdf/?doclanguage=en&cote=sg/sd%282002%291/final. Accessed in June 2012.

Powell, W. W., & Snellman, K. (2004). The knowledge economy. *Annual Review of Sociology, 30*, 199–220.

Ross, M., Larson, E. D., & Williams, R. H. (1987). Energy demand and materials flows in the economy. *Energy, 12*(10/11), 953–967.

Stern, D. (2010). Energy quality. *Ecological Economics, 69*, 1471–1478.

van den Bergh, J. C. J. M. (2010). Relax about GDP growth: Implications for climate and crisis policies. *Journal of Cleaner Production, 18*(6), 540–543.

Warr, B. S., & Ayres, R. U. (2010). Evidence of causality between the quantity and quality of energy consumption and economic growth. *Energy, 35*, 1688–1693.

Ziolkowska, J. R., & Ziolkowski, B. (2011). Product generational dematerialization indicator: A case of crude oil in the global economy. *Energy, 36*, 5925–5934.

Chapter 5
Re-evaluating Energy Intensity: A New Methodological Framework

It remains true, however that the greatest inequalities stem from those who pursue excess not from those driven by necessity. One does not become a tyrant to stop feeling cold.

Aristotle

Abstract This chapter delineates an alternative framework for the empirical evaluation of the link between energy and growth. The fundamental principle is that the outcome of the economic system can only be seen at the human scale. The economy produces goods that provide economic welfare to human beings. Economic welfare is an individualistic perception arising from the satisfaction of human needs. Therefore the outcome of the economic system cannot be approximated by aggregate GDP without a clear reference to the people possessing this outcome. The proposed approach sets the economic process within the Coupled Human and Natural Systems (CHANS). This perception sheds light on two parameters that irrevocably define the Energy Intensity of the economy: the population related to the economy and the biophysical properties that good should have in order to be able to serve human needs.

Keywords Human scale · Coupled human and natural systems · Economic welfare and utility · Biophysical analysis

5.1 The Economic System Within the Coupled Human and Natural Systems (CHANS)

Economies expand intensively by dominating new spheres of contemporary social life. Certain social activities that contribute significantly to social welfare are transformed from non-market to market activities. This shift contributes to an increasing GDP index, further boosting the inherent trends in economic growth. For example, child care, an activity traditionally performed by parents or grandparents, has over the last 20 years become a market activity which increases GDP even though it does not necessarily lead to actual augmentation of the welfare enjoyed by families.

© The Author(s) 2016
K. Bithas and P. Kalimeris, *Revisiting the Energy-Development Link*,
SpringerBriefs in Economics, DOI 10.1007/978-3-319-20732-2_5

The expansion of economic life to spheres of social life promotes a new and probably erroneous vision of the economic process. As the economic process and the economic system expand, they acquire a fundamental "independence" from the rest of social life. The economic system becomes an autonomous system which pursues its own objectives. The economic system is separated from the social system and the economic process is envisaged as being independent from social processes. Within this vision, the economic system achieves economic objectives that are considered to be quite separate from social objectives. The economic system is envisioned as an engine that creates benefits and profits which are estimated in monetary units. The higher the aggregate benefits, the better the economic system. As a result, the aggregate monetary units created by the economic system indicate its effectiveness and are adopted as the sole indicator of the performance of the economy. This simply implies that GDP has gained substantial appeal as being the major indicator utilized for evaluating the economy. That such a vision fails to recognize the ultimate objective of the economic system and the essence of the economic process is brushed aside.

The economic system creates the economic welfare enjoyed by human beings. The satisfaction of certain human needs is the "cause" of the economic process. The economic system, functioning within the social system, attempts to satisfy a certain spectrum of human needs. Economic processes are interwoven with social processes. Human beings are the ultimate actors in both the social and the economic processes. New economic entities such as multinational enterprises and offshore companies reflect the pursuit of economic welfare by the human beings who "own" these economic entities. Indeed, such economic organization reflects the complexity of the contemporary economic world. Nevertheless, the interests of human beings underlie the operation of the economic world.

Economic systems cannot be separated from human social systems. The economic system is an integral part of the human system, one that encompasses the creation of that portion of human welfare that arises from the use of goods exchanged in markets. On the other hand, the economic system is inevitably conditioned by natural laws. The economic process is a biophysical process that obeys the laws which regulate the natural system. The natural elements of the economic process are all of the endowments of the earth's global ecosystem. The economic system is a subsystem of the natural system (Passet 1979; Georgescu-Roegen 1971, 1982, 1986; Daly 1997, 2013) and it inevitably functions within coupled Human and Natural Systems (CHANS) (Liu et al. 2007; Mavrommati et al. 2014). The economic system is a subsystem within the CHANS as depicted graphically in Fig. 5.1 adopted from Passet (1979).

Once the economic system is perceived as the engine that produces the economic utility-welfare which satisfies human beings, then the outcome of the economic system can only be evaluated in monetary terms by the index "GDP per Capita". GDP per Capita is not the only monetary indicator that approximates the ultimate outcome of the economic system—the economic utility enjoyed by human beings. Other measures of human utility/well-being have been formulated in recent years by prominent academics and international organizations. (Max-Neef 1995; Daly and

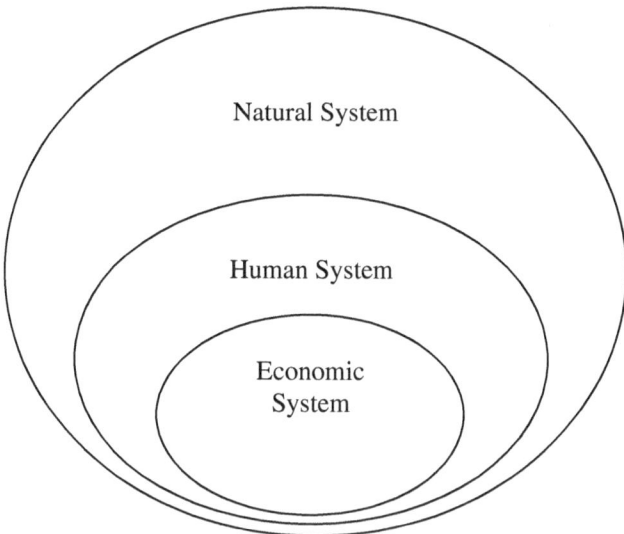

Fig. 5.1 The economic system within CHANS

Cobb 2007; Kubiszewski et al. 2013; Costanza et al. 2014). These innovative indexes avoid the single aggregate monetary framework and extend to multidimensional scales. These indicators take into account several factors which do not yet enjoy widespread scientific acceptance and constitute experimental scientific efforts to approximate real welfare in monetary terms. For the time being, they are applied to a very small number of economies for relatively short periods of time. Therefore they do not yet permit the analysis of macro and structural shifts in the link between economy and energy. As a result, GDP per Capita stands as the only available indicator of economic welfare which is widely used and provides long run data for all economies. It is the standard monetary-based indicator for economic welfare which, although incomplete, permits macro and structural analysis. Furthermore, GDP per Capita is based on the relevant preferences taking place in actual markets, and therefore reflects actual and "objective" economic reality. Within the monetary context, the index of GDP per Capita emerges as the only monetary indicator of welfare/utility with long-run data availability for the majority of national economies, allowing us to perform world-wide and long run EI analysis.

5.2 GDP as a Proxy of "GDP Per Capita": An Analytical Assumption

Economic analyses, reports and discussions are often based on the trends in GDP as the measure of the state of the economy. Behind this practice, there lies a simplified analytical assumption that concerns the broad use of aggregate GDP as a measure of the economic output. Within a socio-economic system, the number of human beings, the population size, changes at extremely slow rates. Population trends are infinitesimal, relative to trends in economic output measured by GDP. Population size evolves smoothly in periods of peace whereas the economy grows at relatively intensive rates. For example, the US and Japanese populations increased by 86 and 51 %, respectively, from 1950–2000, while aggregate GDP increased by 460 and 1513 %, respectively, over the same period (The Conference Board Total Economy Database 2014).

Within this context, the "Capita", of the index "GDP per Capita" evolves at much slower rates than GDP. As a result, trends in GDP adequately approximate the trends of GDP per Capita, especially for short-run analysis. With the population almost constant between two successive years, a 5 % growth rate of GDP is almost equal to a 5 % growth in GDP per Capita. This fact fostered the use of GDP as an approximation to GDP per Capita. In the short run, when the population size changes minimally, the growth of GDP implies a similar increase in the utility enjoyed by human beings, as estimated in monetary terms by GDP per Capita.

As GDP trends closely approximate the trends of GDP per Capita, GDP has implicitly replaced GDP per Capita as a measure of the performance of the economic system in various economic studies and reports. This fact has induced an erroneous view of the use of GDP as the appropriate index for the evaluation of the performance of the economy and hence the long run changes in the same economy and, subsequently, to facilitate comparisons between different economies. Nevertheless, GDP cannot sufficiently approximate structural changes that occur over the long run evolution of an economic system, nor the different structures of different economies. To cite an indicative example, trends in the GDP of the USA from 1950–2000 cannot reflect the actual changes in the output of the US economy. These can only be approximated, in monetary terms, by the trends in the GDP per Capita index. Furthermore, GDP cannot be used for comparing the economies of the USA and China which share a similar level of GDP. Although the USA (10,036,395 million 1990 GK$, in 2012) and China (11,752,373 million 1990 GK$ in 2012) share similar GDPs, they represent two economies with extremely different levels of effectiveness, as they provide substantially different levels of utility to their citizens. The differences in the utility level enjoyed by the citizens of the USA and China, are clearly reflected in the substantially different levels of their GDP per Capita with values of 31,913 in the USA and 8698 1990 GK$ in China in 2012. Behind the different levels of utility, there exist two completely different economic systems with different economic processes and structures.

To conclude, although trends in GDP have been adopted as a proxy for trends in GDP per Capita when long run changes are examined, aggregate GDP cannot be used as a fundamental measure of the outcome of an economic system. GDP is a misleading indicator for long run, structural analysis and international comparisons of national economies. The use of GDP as the monetary index of the performance of an economy brushes aside the very essence of the economy which is the creation of utility and which can only be approximated in monetary terms by the GDP per Capita index.

5.3 Energy Intensity Evaluated at the Borders of the Economic System

Energy is an indispensable input to the economic process. It provides the power for the production process in which capital and labour process natural resources in order to create goods. Capital, in order to function, requires energy inputs (Georgescu-Roegen 1971, 1975; Daly 1997). Energy could be considered the engine of growth at the macro level of the economy, since energy is the indispensable fuel of the production process at the micro level (Ayres and Warr 2009; Ayres et al. 2013). The economic process utilizes inputs, factors of production, in order to create goods. Goods are the useful outcome of the production process. Human beings enjoy the utility/welfare arising from the consumption of goods. The creation of welfare is the ultimate outcome of the production process and, therefore, of the economic system. Goods are not abstract material structures but they embody those biophysical properties which satisfy human needs and hence create welfare. Goods could be perceived as the "means" for the satisfaction of needs and hence for the enjoyment of welfare-utility. Figure 5.2 offers a simplified representation of the physiology of the economic system.

The boundaries of the economic process cannot be drawn before the production of goods has been completed; therefore they are drawn to reflect the very essence of the economic process, which is the creation of welfare. The micro-economic process and the macro-economic system can only be examined as functions that make use of natural resources (energy and mass inputs) in order to produce goods that satisfy human wants and create human welfare.

The effectiveness of the economic system to utilize energy inputs should be evaluated at the "borders" of the economic system. Energy inputs enter the economic system, and goods that provide welfare to human beings are the outcome of the system. The energy effectiveness of the system is estimated as the energy inputs

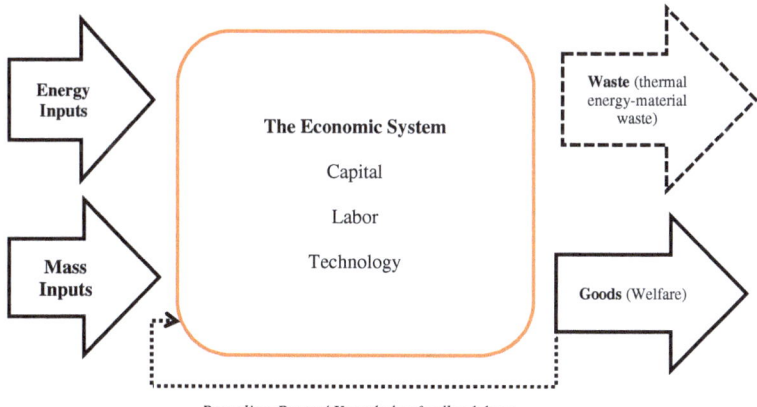

Recycling-Reuse / Knowledge feedback loop

Fig. 5.2 The economic system

required to create welfare. The energy requirements of economic welfare define the Energy Intensity (EI) of the economic system and emerge as an essential bio-physical property of the economy.

The economic welfare/utility enjoyed by human beings is estimated through the monetary index of GDP per Capita which emerges as the monetary index of the ultimate output of the economic system. Within the monetary domain, EI is then evaluated by relating the energy inputs entering the economic system with its actual outcome, measured as GDP per Capita. In effect, EI is evaluated at the "borders" of the economic system by comparing those flows entering into the economic system with those resulting from it. Hence, EI is estimated by the *Energy/[GDP per Capita]* Index. With GDP per Capita indicating the monetary index of welfare/utility, EI can be defined as *Energy/Welfare* or *Energy/Utility*. Thus, EI indicates the energy inputs required for the creation of one unit of economic welfare/utility.

5.4   EI and the Biophysical Properties of Commodities

Energy Intensity (EI) is a fundamental property of the economic system, co-determined by the biophysical properties of the goods produced, as well as by the technological status of the economic process. EI is estimated within the monetary context by relating energy inputs to the outcome of the economic system which is estimated in monetary units. Monetary units are the converters that allow numerous different goods to be compared on a scale which reflects economic welfare. Monetary units function as a veil that "covers" and "homogenizes" all the goods produced; they are a prism that projects actual goods onto a common scale, reflecting economic welfare/utility. Underneath the veil of monetary units there exist actual goods with certain properties. Although EI is a monetary-based index, relating energy with monetary units, its value is determined by the biophysical properties of the goods produced as well as by the technological status of the production process. Evidently, the very same aggregate monetary values (GDP) and energy inputs may result in completely different EIs, as different sets of goods may correspond to the same aggregate GDP. EI, being among the fundamental bio-physical properties of the economy, ought to be evaluated only with clear reference to the biophysical characteristics of the actual goods produced. On the other hand, a macro index, such as EI, cannot reflect the properties of each and every individual good. EI is an aggregate indicator which should be estimated at the appropriate level of aggregation in the economy. EI ought to be evaluated at an aggregation level that is sufficiently "macro" to reflect the overall structure of the economic system which encompasses all individual production processes. At the same time, the appropriate macro level of aggregation should not fail to reflect the biophysical properties of the actual goods. GDP emerges as the highest level of monetary aggregation summing up the monetary values of all goods. GDP is a monetary amalgam. At the level of aggregate GDP, all goods have been homogenized in abstract monetary units and the implications arising from their physical properties have been eliminated from the analytical picture.

Downscaling GDP to the monetary level of GDP per Capita makes the implication of the physical properties of goods discernible, albeit indirectly (Fig. 5.3). GDP per Capita, the so-called "income" index, reflects the monetary value of the set of goods that is consumed by the average citizen. This set of goods is a satisfactory representation of the actual goods produced by the economy. The economic meaning of the GDP per Capita index is that it corresponds to the monetary value of the representative bundle or basket of goods consumed by the average citizen. This bundle or basket consists of a mixture of all goods (food, housing, transport, services, etc.) produced within the economy. Notably, these goods are produced at different rates in each economy. This means that these economies differ and hence have different EIs. The shares of the different economic sectors cannot be reflected in the aggregate GDP whereas they are indirectly approximated by the GDP per Capita index. Indeed, the contributions of different economic sectors are indirectly reflected in the representative basket of goods consumed by the average citizen. Let us compare two economies with the same aggregate GDP but substantially different population sizes, which results in substantially different GDPs per Capita. The economy with the higher GDP per Capita reflects a relatively higher production of service-like goods. In effect, when the biophysical properties of actual production are to be determined, the monetary index of per Capita GDP presents significant advances relative to aggregate GDP.

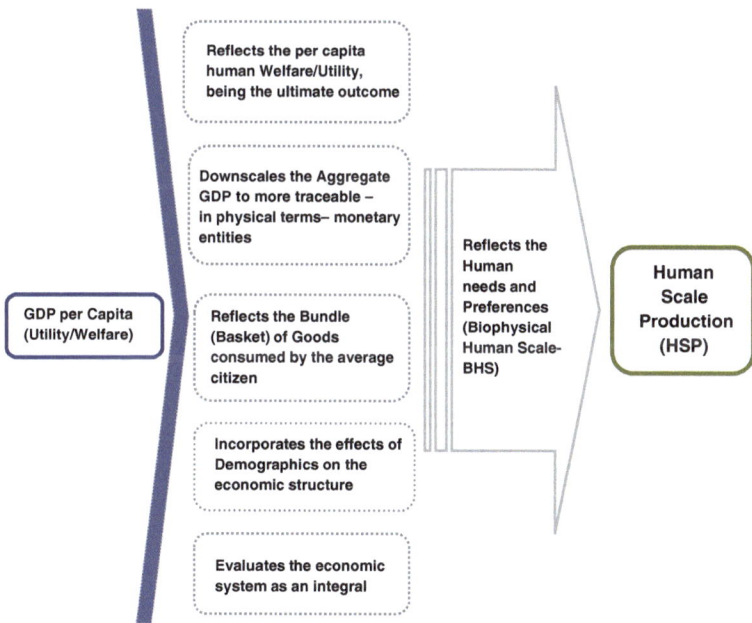

Fig. 5.3 Human scale of the production process

The representative synthesis of goods included in GDP per Capita makes their physical properties definable, even in an indirect way, since GDP per Capita introduces into the analysis the underlying reason for the economic process. The satisfaction of human needs and wants is the reason, the cause, of the production process (Fig. 5.3). Goods should have certain biophysical properties in order to satisfy human needs, i.e., food should offer certain vitamins and a quantity of calories; shelter should have certain minimum dimensions; transport vehicles should also have certain dimensions and other physical attributes to transport people. Goods should be endowed with certain biophysical characteristics defined by the nature of human beings. Goods, in order to be useful, should be on a "human scale" (Max-Neef 1991, 1992; Folke et al. 1996; Fogel 1999; Gibson et al. 2007; Cruz et al. 2009). GDP per Capita indicates, in monetary terms, the human scale of production by downscaling aggregate GDP to the per capita, human, level. On the contrary, aggregate GDP is deprived of any reference to the human scale (Gowdy 1997; Daly 1997, 2013; Fogel 1999; Lawn 2001; Costanza et al. 2014).

The "human scale" of production implies that goods should have certain bio-physical properties, the most important of which are the following:

- Scaffolding, physical dimensions.
- Embodied energy.

Scaffolding reflects the necessary dimensions of goods that are determined by the dimensions of the human body as well as by the dimensions of the common exosomatic instruments related to human life (Georgescu-Roegen 1971, 1977). Houses cannot be less than a certain number of square meters in order to meet the needs of human beings. Although the limit may not be definable beyond dispute, a house of 3 m^2 cannot meet the needs of an individual, even less those of a family. Similarly, a car of matchbox size cannot be functional. A hammer, a common exosomatic instrument, strongly related to human evolution, cannot be useful if it is below a certain weight. And the storage of all "hammer-like" instruments and tools necessitates a house of more extensive dimensions than a house which exclusively facilitates a human body. Inalienably, goods are embodied in certain physical dimensions, scaffolding, which cannot shrink below a certain level. The creation of the scaffolding of goods during the production process makes the use of mass resources necessary. This necessity defines the material intensity of production. Inevitably the processing of mass resources requires energy inputs. Energy is the "fuel" of the production process. Capital and labour process mass inputs with the power provided by energy inputs. Hence, there is causality stemming from the nature of human needs to the energy requirements of the production process. In addition, the most telling example is offered by the direct biological energy requirements of human beings which can only be satisfied through the consumption of at least a minimum number of calories. Food is necessary for human survival and provides the energy inputs which are necessary for human biological life. Therefore, food should have at least a minimum EI.

The nature of human needs sets limits for the material and energy requirements of the economy. The actual mass and energy inputs in each historical period are conditional upon the technological status of the production. Although technology matters, the nature of human beings nevertheless defines certain limits that cannot be overcome by technology. The power of technology cannot overcome the limits set by the biophysical properties of human beings.

Once we accept that the Energy Intensity (EI) of the economy is irrevocably driven by the nature of human beings, the evaluation of EI should take into account the implications set by the nature of human needs. Although EI is a monetary-based index which relates energy flows with the monetary value of the outcome of the economic system, EI should be evaluated for a monetary entity that reflects human beings as the "cause" of the economic process and, therefore, as a restriction of the EI potentials.

The GDP per Capita index is the only monetary-based index for the outcome of the economic system that is defined in reference to the "reason" behind the economic process, human beings. As a result, GDP per Capita emerges as the appropriate aggregate monetary measurement for the evaluation of the EI of the economic process. It provides the necessary link between abstract monetary units and the biophysical properties of goods determined by the human scale. The energy inputs required for the production of one unit of GDP per Capita indicate the energy intensity of the average outcome of the economy. As the "representative" set of goods comprising the basket reflected by the GDP per Capita index is a satisfactory approximation of the actual structure of production, the EI of GDP per Capita provides a good approximation to the average energy intensity of the economy. On the other hand, the evaluation of energy intensity at the level of aggregate GDP estimates the energy intensity of an abstract monetary unit, i.e. the energy inputs consumed for the production of one unit of GDP. Aggregate GDP contains no reference to actual goods, just to their monetary values. GDP is a homogenous monetary amalgam. However, human beings do not consume monetary units. Human wants and needs are satisfied by actual goods. As a result, the evaluation of EI should take place at a monetary level that reflects the human scale of production (HSP in Fig. 5.3) and the respective implications for energy requirements. The "cause" of production cannot be excluded from a robust evaluation framework.

To conclude, EI is among the fundamental biophysical characteristics of the economy. Although EI is a monetary-based index, it can only be evaluated at an aggregate monetary level whose biophysical properties are traceable. GDP per Capita, defined as the monetary value of goods consumed by the average citizen, approximates the average synthesis of the goods produced within an economy. The energy inputs required for the production of one unit of GDP per Capita offers an evaluation of the EI that approximates the actual structure of the economy, the relative share of the different economic sectors and the biophysical properties of goods. On the other hand, the prevailing EI indicators, based on the energy inputs required for one unit of GDP, ignore the structure and the synthesis of production as well as its biophysical properties that determine EI. With GDP being the monetary amalgam of the output of the economy, the Energy/GDP index offers a sallow evaluation of EI which may result in misleading results since the biophysical coordinates of actual production are completely ignored.

5.5 Population Size, GDP, and Energy Intensity

Economies of the same aggregate GDP may serve the needs of vastly different numbers of human beings. The difference in population size makes economies different in several fundamental ways, with resource requirements being crucial among these.

Population size is linked directly to the number and the structure of human needs that seek satisfaction in the ultimate outcome of the economy. A larger population, ceteris paribus, implies a higher percentage of "basic" needs such as food, shelter, heating. On the contrary, an economy with a smaller population, ceteris paribus, is oriented towards a higher percentage of service-like and luxury goods. Logic dictates that population size matters for the structure of production.

As population size changes across nations and over time, economies of similar aggregate GDP may produce fundamentally different categories of goods. Nevertheless the influence of population size cannot be reflected in aggregate GDP, since GDP is insensitive to population size. On the other hand, the monetary index of GDP per Capita approximates the effects of the population on the structure of production more effectively. That is why it is broadly accepted that economies with higher GDP per Capita, such as those cited as developed economies, are oriented more intensively towards services and luxury goods.

The structure of production, i.e. the relative shares of economic sectors, has direct implications for the Energy Intensity of the economy. As the structure of production is better approximated by GDP per Capita, this index emerges as the appropriate monetary level for the evaluation of the energy requirements of the economy.

The insistent causality between population size and the structure of production has direct implications for the energy intensity of production. In effect, only a monetary based indicator that is sensitive to population size can be used as the appropriate basis for the evaluation of Energy Intensity.

5.6 EI and the Energy Metabolism of the Socio-Economic System

The "standard" empirical estimates of the EI of several national economies demonstrate a declining trend which indicates a gradual delinking between energy and the economy (see Chap. 4). These estimates are all based on the standard EI indicator which is defined as *"the energy inputs required for the production of one unit of GDP*": Domestic Energy Consumption (DEC[1])/GDP. On the other hand, the energy used by each human being, defined as "DEC per capita", demonstrates a dramatically increasing trend. DEC per capita is adopted as a measure of the so-called energy metabolism of the socio-economic system. Both indicators take into account the very same energy inputs i.e. the amount of energy that is used by the economic system in the course of the production process which creates all the domestically produced goods within an economic system. Indeed, the energy use reflected in both indices, DEC/GDP and DEC per capita, reflects the very same energy inputs utilized by the economic system.

The drastically increasing difference between DEC/GDP and DEC per capita raises important issues for the appropriateness of the prevailing framework for evaluating the link between energy and the economy. A first intuitive explanation of the increasing difference may be traced to the disproportionate increase in the GDP per Capita index. Indeed, in the very same periods, the growth of GDP per Capita implies that the average citizen consumes more goods, as well as goods of higher monetary value, and therefore enjoys higher levels of utility. Evidently, the trends in utility enter into the analysis as the explanatory factor for a fundamental contradiction, the increasing difference between DEC/GDP and DEC per capita, which has been noted within the prevailing evaluation framework. Indeed, the divergence between these two indicators that both estimate the energy requirements of the socio-economic system can only be explained with reference to the monetary index of GDP per Capita approximating the ultimate objective of the economic system, the satisfaction of human needs, and thus the utility level. Curiously, although the trends in utility are used as the explanatory reason which resolves the contradiction, utility has been completely ignored in the prevailing EI estimates.

The actual link between energy and the economy cannot be evaluated without taking into account the actual outcome of the economy, economic utility, approximated in monetary terms by GDP per Capita. As a result, the indicator "*energy input required for the production of one unit of GDP per Capita*" is adopted as the appropriate index of the EI. It is expected that the difference between DEC/Utility and DEC per capita, will be substantially lower than the difference between DEC/GDP and DEC per capita. If this is indeed so, the relevant estimates can be interpreted as proof of the superiority of DEC/Utility over DEC/GDP. Indeed, the EI of the economy cannot follow a substantially different trend from the energy metabolism of the socio-economic system. The social metabolism indicates the EI of one human being in the pursuit of its utility, while the EI of the economy evaluates the energy requirements of one unit of utility created by the economic system. The energy requirements of the utility enjoyed by the average citizen cannot exhibit substantially different trends from the energy requirements of one unit of utility.

---

[1]A detailed definition of DEC is provided in Sect. 6.1.

5.7 Constraints and Limitations

The use of GDP per Capita as an approximation to the actual outcome of the economic system is based on a number of assumptions. Although these assumptions are common in economic analysis, it is worth mentioning them here in order to make explicit the analytical framework adopted for the evaluation of EI. GDP per Capita is envisaged as the monetary value of the set of goods consumed by the average citizen. This set of goods reflects the representative basket of goods used by human beings in each socio-economic system. This representative basket consists of a combination of goods that reflects the relative shares of all the different production sectors and industries within an economy. It consists of food, shelter, transport, services and other goods, in the relative shares in which they are produced by the economic system. Hence, the representative set of goods defined by GDP per Capita offers an approximation of the average synthesis of all different goods produced by an economy. However, this representative basket of goods is an analytical instrument which does not exist in reality. Not all people consume the very same goods; all the more, the synthesis of every individual's consumption does not correspond to the relative shares of the different industries of an economy. As a result, GDP per Capita should be seen as an analytical entity to approximate the actual synthesis of different goods produced within an economy.

Telling evidence concerning the essential economic meaning of the GDP per Capita index is offered by the classification of the world economies into developed and developing ones. The broadly accepted distinction between developed and developing economies does reflect, albeit indirectly, a difference in the relative shares of the economic sectors which is indicated in monetary terms by the different levels of the GDP per Capita (income) index. Indeed, economies with the same aggregate GDP, but substantially different population sizes, reflect different relative shares of their industries. An economy with higher per Capita GDP is expected to be more heavily oriented towards service-like goods. This fact cannot be depicted in aggregate GDP.

Notably, GDP per Capita is completely insensitive to the distribution of income among individuals. Distribution of income may result in substantial effects on the synthesis of production and therefore on the energy requirements of the economy. These effects cannot be captured by the GDP per Capita index, which estimates the average consumption of the average citizen, which is a theoretical edifice serving analytical purposes.

Indeed, GDP per Capita, as an index that approximates the outcome of the economic system, is burdened with certain assumptions which often result in a simplified representation of reality. However the real question should concern the comparison between the two monetary-based indices of the economy, GDP and GDP per Capita. The latter offers certain advantages over the former as an approximation of what is really going on within the economic system. GDP per Capita emerges as an indicator that reflects certain characteristics of the economy. By introducing the "cause" of the economic process into the analysis, this indicator reflects the human scale of production (HSP in Fig. 5.3) and therefore emerges as a monetary index that can be used for tracing the biophysical physiology of the goods produced. Within this context, GDP per Capita has clear advantages over the monetary amalgam that the GDP index is.

5.8 The Operational Structure of the Energy-Economy Indicators

The evaluation of the link between energy and the economy will be based on the Domestic Energy Consumption (DEC)/Utility ratio. This is the core indicator for the proposed approach of the present volume and its properties have been systematically presented in the previous sections. Inevitably, the trend of the core indicator will be compared with the respective trends of the standard DEC/GDP ratio. As these indicators consist of different algebraic structures which result in different scales, the interest of the analysis lies in the comparison of the trajectories of the two indicators and, especially, in the directions of their trends. The two indicators, DEC/Utility and DEC/GDP, have been indexed to a base year, namely t_0, where:

$$\text{Indexed value}\, t = 1 + (t_i - t_0)/t_0,$$

$$\text{with } i = \text{a specific year within the period examined} \qquad (5.1)$$

By scaling both indicators to a base year (indexed $t_0 = 1$), we eliminate scale differences arising from their different absolute values. The value of both indicators in the base year is the same (5.1) and their trends are estimated on an annual basis.

Beyond the two indicators, we estimate a number of additional indices that reveal further important aspects of the Energy-Economy link. First, the per capita use of energy (DEC per capita) is estimated as an indicator that reveals the so-called social/societal (or industrial) metabolism (Ayres 1989; Baccini and Brunner 1991; Ayres and Simonis 1993; Fischer-Kowalski and Hüttler 1998). Second, the Decoupling Index (DI), as proposed by the United Nations Environment Report for Decoupling (UNEP 2011), is estimated. In its original version, the Decoupling Index refers to the ratio of the change in the rate of consumption of energy, to the change in the rate of economic growth (in terms of GDP), on an annual basis. In order to smooth out short-term fluctuations in economic cycles we estimate DI for each decade, by using moving averages instead of annual estimates.

We estimate the DI for the DEC/GDP ratio as:

$$\begin{aligned} DI &= \Delta(\text{DEC})/\Delta(\text{GDP}) \\ &= [(\text{DEC}_t - \text{DEC}_{t-1})/\text{DEC}_{t-1}]/[(\text{GDP}_t - \text{GDP}_{t-1})/\text{GDP}_{t-1}] \qquad (5.2) \end{aligned}$$

where t is an averaged time period of one decade. We estimate the DI for DEC/Utility indicator in the same way. Fisher-Kowalski et al. (UNEP 2011) propose the following interpretation for DI:

- When DI > 1, no decoupling is taking place.
- When DI = 1, the decoupling threshold is being approached.
- When 0 < DI < 1, relative decoupling is taking place.
- When DI = 0, it is implied that the economy is growing while resource consumption remains constant. This is the turning point between relative and absolute decoupling.
- When DI < 0, absolute decoupling has been achieved.

The Difference between Energy Intensity (EI) and social/industrial metabolism

We estimate and compare the trends of per capita DEC with the EI trends as estimated through the DEC/GDP and DEC/Utility indicators. The trends in the percentage (%) change in the difference between DEC per capita and DEC/GDP, and between DEC per capita and DEC/Utility, are assessed and compared as follows:

$$\text{Trends in } \% \text{ change } t = [(\text{indexed DEC per capita})t - (\text{indexed DEC/GDP})t]\% \tag{5.3}$$

$$\text{Trends in } \% \text{ change } t = [(\text{indexed DEC per capita})t - (\text{indexed DEC/Utility})t]\% \tag{5.4}$$

Both of these functions (5.3–5.4) are estimated only for the cases with long run data, namely for the global economy, the USA, and Japan.

References

Ayres, R. U. (1989). Industrial metabolism. In J. H. Ausukl & H. E. Sladovich (Eds.), *Technology and environment* (pp. 23–49). Washington, D.C.: National Academy Press.

Ayres, R. U., & Simonis, U. E. (1993). *Restructuring for sustainable development*. University of Nogura, Department of Natural Resources Publication no. 35. Tokyo: United Nations University Press.

Ayres, R. U., Van den Bergh, J. C., Lindenberger, D., & Warr, B. (2013). The underestimated contribution of energy to economic growth. *Structural Change and Economic Dynamics, 27*, 79–88.

Ayres, R. U., & Warr, B. (2009). *The economic growth engine: how energy and work drive material prosperity*. UK: Edward Elgar Publishing.

Baccini, P., & Brunner, P. H. (1991). *Metabolism of the anthroposphere*. Berlin/Heidelberg: Springer.

Costanza, R., et al. (2014). Time to leave GDP behind. *Nature, 505*, 283–285.

Cruz, I., Stahel, A., & Max-Neef, M. (2009). Towards a systemic development approach: Building on the Human-Scale Development paradigm. *Ecological Economics, 68*, 2021–2030.

Daly, H. E. (1997). Georgescu-Roegen versus Solow/Stiglitz. *Ecological Economics, 22*(3), 261–266.

Daly, H. E. (2013). A further critique of growth economics. *Ecological Economics, 88*, 20–24.

Daly, H. E., & Cobb, J. B, Jr. (2007). ISEW: The 'debunking' interpretation and the person-in-community paradox: Comment on Rafael Ziegler. *Environmental Values, 16*(3), 287–288.

Fischer-Kowalski, M., & Hüttler, W., (1998). Society's metabolism. Journal of Industrial Ecology, *2*(4), 107–136.

Fogel, R. W. (1999). Catching up with the economy. *American Economic Review, 89*(1), 1–21.

Folke, C., Holling, C. S., & Perrings, C. (1996). Biological diversity, ecosystems, and the human scale. *Ecological Applications, 6*(4), 1018–1024.

Georgescu-Roegen, N. (1971). *The entropy law and the economic process.* Cambridge: Mass.

Georgescu-Roegen, N. (1975). Energy and economic myths. *Southern Economic Journal, 41*(3), 347–381.

Georgescu-Roegen, N. (1977). Inequality, limits and growth from a bioeconomic viewpoint. *Review of Social Economy, 35*(3), 361–375.

Georgescu-Roegen, N. (1982). Energetic dogma, energetic economics, and viable technologies. *Advances in the Economics of Energy and Resources, 4*, 1–39.

Georgescu-Roegen, N. (1986). The entropy law and the economic process in retrospect. *Eastern Economic Journal, 12*(1), 3–25.

Gibson, C. C., Ostrom, E., & Ahn, T. K. (2007). The concept of scale and the human dimensions of global change: A survey. *Ecological Economics, 2*, 217–239.

Gowdy, J. M. (1997). The value of biodiversity: Markets, society, and ecosystems. *Land Economics, 73*(1), 25–41.

Kubiszewski, I., Costanza, R., Franco, C., Lawn, P., Talberth, J., Jackson, T., & Aylmer, C. (2013). Beyond GDP: Measuring and achieving global genuine progress. *Ecological Economics, 93*, 57–68.

Lawn, P. A. (2001). Goods and services and the dematerialisation fallacy: Implications for sustainable development indicators and policy. *International Journal of Services, Technology and Management, 2*(3), 363–376.

Liu, J. G., Dietz, T., Carpenter, S. R., Alberti, M., Folke, C., Moran, E., et al. (2007). Complexity of coupled human and natural systems. *Science, 317*, 1513–1516.

Mavrommati, G., Baustian, M. M., & Dreelin, E. A. (2014). Coupling socioeconomic and lake systems for sustainability: A conceptual analysis using Lake St. Clair region as a case study. *Ambio, 43*(3), 275–287.

Max-Neef, M. A. (1991). *Human scale development: Conception, application and further reflections.* NY: The Apex Press. Available at: http://espace.library.uq.edu.au/view/UQ:340489 . Accessed February 2012.

Max-Neef, M. A. (1992). *From the outside looking in: Experiences in 'barefoot economics'.* London, UK: Zed Books Ltd.

Max-Neef, M. (1995). Economic growth and quality of life: A threshold hypothesis. *Ecological Economics, 15*(2), 115–118.

Passet, R. (1979). *L e'conomique et le vivant.* Paris: Payot.

The Conference Board Total Economy Database. (2014). Available at: http://www.conference-board.org/data/economydatabase/.

UNEP. (2011). *Decoupling natural resource use and environmental impacts from economic growth.* Fischer-Kowalski, M., Swilling, M., von Weizsäcker, E. U., Ren, Y., Moriguchi, Y., & Crane, W., et al. A report of the Working Group on Decoupling to the International Resource Panel, United Nations.

Chapter 6
Empirical Analysis

For the end of economy is not the physical augmentation of goods but always the fullest possible satisfaction of human needs.

Carl Menger, Principles of Economics, 1871: p. 190

Abstract The Energy Intensities of 22 countries as well as the global economy are estimated on the basis of the framework proposed in Chap. 5. The estimates are compared with those based on the prevailing framework as presented in Chap. 4. The differences are essential and imply completely different perspectives for the dependency of future growth on energy. In order to shed light on the intricate link between energy and growth, estimates for a number of additional indicators are also provided.

Keywords Energy intensity · Social metabolism · Decoupling index · Developed and developing economies

6.1 Data Overview

Two distinct datasets on energy consumption feed the empirical estimates of the present book:

1. All data on energy use concerning global economy, USA and Japan are drawn from the databases of the Institute of Social Ecology of Alpen-Adria University in Vienna, Austria (available online at: http://www.uni-klu.ac.at/socec/inhalt/ 1088.htm). Specifically, the data for the global energy supply are drawn from Krausmann et al. (2009) covering the period 1900–2009; Domestic Energy Consumption (DEC) for the USA are drawn from Gierlinger and Krausmann (2012) covering the period 1870–2005; DEC for Japan are drawn from Krausmann et al. (2011) covering the period 1878–2005. For the global level, all data on Total Primary Energy Supply (TPES)[1] are expressed in Exa-joules per

---

[1]The total primary energy supply (TPES) is the sum of all energy resources worldwide. Consequently, TPES could be perceived as the total sum of each national DEC of all the countries of the world.

© The Author(s) 2016

K. Bithas and P. Kalimeris, *Revisiting the Energy-Development Link*,
SpringerBriefs in Economics, DOI 10.1007/978-3-319-20732-2_6

year (EJ/year). For the cases of the USA and Japan, all data on DEC are expressed in Peta-joules per year (PJ/year).

2. BP Statistical Review of World Energy (2014): Data on DEC for the cases of Australia; Canada; Germany; Italy; the Netherlands; Norway; South Korea; Spain; Taiwan; the United Kingdom; Argentina; Brazil; China; India; Indonesia; Malaysia; Mexico; Pakistan; Thailand; and Turkey, as well as the energy data utilized for the short run estimates (Sect. 6.5) of the USA and Japan, are drawn from the BP Statistical Review of World Energy (Available online at: http://www.bp.com/statisticalreview). BP data cover the period 1965–2013 and are expressed in millions of tons of oil equivalent per year (mtoe/year).

It should be noted that Domestic Energy Consumption (DEC) is estimated in different ways within the two datasets utilized. In the datasets of the Institute of Social Ecology, DEC is estimated in the context of the Economy-Wide Material Flow Analysis (MFA) framework and is defined as "DEC = Domestic Extraction (DE)—exports + imports". The BP dataset calculates domestic energy consumption as "Inland demand plus international aviation and marine bunkers and refinery fuel and losses". For the sake of simplicity, we adopt the term Domestic Energy Consumption (DEC) in the estimates employed in the present volume despite these differences between the two utilized datasets. Although differences may exist between the two methods, the in-depth investigation of this issue remains beyond the scope and purposes of the present book. However, it should be noted that the issues raised by the present book are not affected by the differences between the two datasets.

All data on GDP and population are drawn from Maddison (2008) (available online at: http://www.ggdc.net/MADDISON/oriindex.htm) for the period 1870–2008, while the period 2009–2013 is covered by The Conference Board Total Economy Database (2014) (available online at: http://www.conference-board.org/). GDP is estimated in millions of 1990 International Geary-Khamis dollars per year (million 1990 GK\$/year), and GDP per Capita is estimated in 1990 International Geary-Khamis dollars per year (1990 GK\$/year). The Geary-Khamis dollar is a hypothetical unit of currency that has the same purchasing power parity (PPP) that the United States dollar had in the USA in 1990. It was proposed by Roy C. Geary in 1958 and was further developed by Salem H. Khamis (Geary 1958; Khamis 1972). Population is expressed in thousand persons.

6.2   The Energy Intensity of the Global Economy for 1900–2009

6.2.1   The Trends of Global Primary Energy Consumption Per Energy Type

The world's Total Primary Energy Supply (TPES) increased by 1085.75 % during 1900–2009 (Table 6.1). The aftermath of the WWII is characterized by an excessive

Table 6.1 Percentage change of the major indicators for indicative periods (global)

	1900–2009 (%)	1900–1950 (%)	1950–1995 (%)	1995–2009 (%)
TPES/GDP	−53.79	−14.56	−33.84	−18.25
TPES/Utility	101.48	38.00	50.00	−2.67
TPES	1,085.75	130.99	283.64	33.81
TPES per capita	171.95	43.01	69.20	12.38
GDP	2,466.11	170.35	479.89	63.68
GDP per capita	488.53	67.38	155.76	37.48
Population	336.02	61.52	126.74	19.06

consumption of fossil fuels at the aggregate global level. By 1960, oil consumption is by far the most predominant energy resource being utilized (indicatively 164.6 EJ in 2009), followed by coal (146.5 EJ, in 2009) and natural gas (107.2 EJ, in 2009) (Fig. 6.1a). Biofuel use increases smoothly during 1900–2000, with a more intensive increase during 2001–2009, and concluding at the level of 55.2 EJ in 2009. Furthermore, electricity produced by hydro, geothermal and nuclear sources increases rapidly after 1975, although it remains a very small proportion (43.1 EJ in 2009) of the world's total energy consumption (516.8 EJ in 2009—Fig. 6.1a).

6.2.2   Global Renewable and Non-renewable Energy Consumption

Figure 6.1b depicts the global consumption of renewable and non-renewable energy resources for the period 1900–2009. We aggregate biofuel consumption as renewable energy. Clearly, after 1945, the vast majority of energy resources, at the global aggregate level, are non-renewables (418 EJ in 2009, in contrast with 55.2 EJ of renewables—Fig. 6.1b), demonstrating the heavy dependence of the world's economy on non-renewable energy.

6.2.3   Energy Intensity of the Global Economy

Figure 6.1c presents the Energy Intensity (EI) trends for both the TPES/GDP and TPES/Utility indicators, and the per capita TPES consumption, for the period 1900–2009 (all indexed to the base year 1900, in which all indicators take value = 1). The period 1900–1945, which includes WWI (1914–1918), the Great Depression of 1929, and WWII (1939–1945), contains notable fluctuations, as depicted by both TPES/GDP and TPES/Utility. The year 1951 forms a landmark for both EI

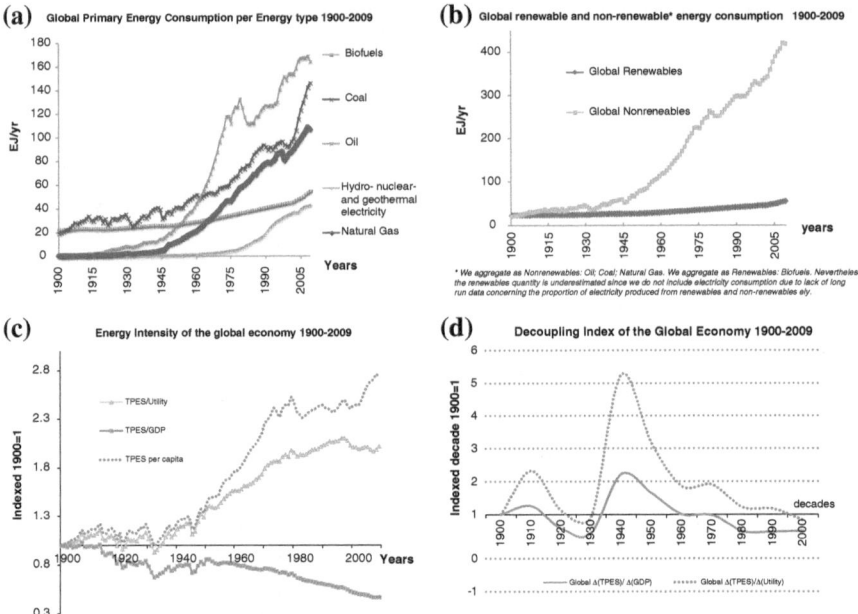

Fig. 6.1 The global energy profile

indicators. The TPES/GDP ratio peaks and, thereafter, steadily declines until 2009. In contrast, the TPES/Utility indicator initiates a strong linkage between energy use and welfare until 1996, stabilizing its trends during 2000–2009. Notably, TPES/Utility increases by 101.48 %, during 1900–2009, a performance similar to the increasing evolutionary path of the per capita TPES (171.95 % during 1900–2009—Table 6.1). On the other hand, the TPES/GDP indicator evolves, in stark contrast, with a reduction by of −53.79 %, over the same period (Table 6.1).

6.2.4   The Decoupling Index (DI) for the Global Economy

This section estimates the Decoupling Index (DI) of the global economy, for the period 1900–2009 (Fig. 6.1d). For the decades of the 1920s and 30s (1920–1939), the DI of the TPES/GDP ratio shows signs of relative decoupling, taking its lowest DI(TPES/GDP) value (0.41) during 1930–1939. For the next three decades, DI (TPES/GDP) takes values ≥ 1 marking a coupling period. DI(TPES/GDP) drops below one after 1970 and until 2009, indicating global economic growth in a state of relative decoupling from TPES. On the other hand, there is no clear sign of any relative decoupling at any time within the entire century (1900–2000) according to the DI(TPES/Utility). An indication of relative decoupling appears only in the period 1930–1939, with DI(TPES/Utility) = 0.9, a value sufficiently close to 1, which marks the "borderline" between coupling and relative decoupling. Notably, the lowest DI(TPES/Utility) value (0.82) occurred in the 2000–2009 period, which possibly indicates the start of a relative decoupling period (Fig. 6.1d).

6.2.5   The Difference Between the Per Capita Energy Consumption and Energy Intensity Trends for the Global Economy

Figure 6.2 presents the % change of the difference between the indexed values of TPES per capita and: (1) TPES/GDP; and (2) TPES/Utility. Concerning case (1), a continuously increasing difference is observed throughout the period under examination. On the other hand, case (2) depicts almost no difference during 1900–1950, while the period 1950–2009 reveals an increasing difference, albeit smoother than the corresponding one observed in case (1).

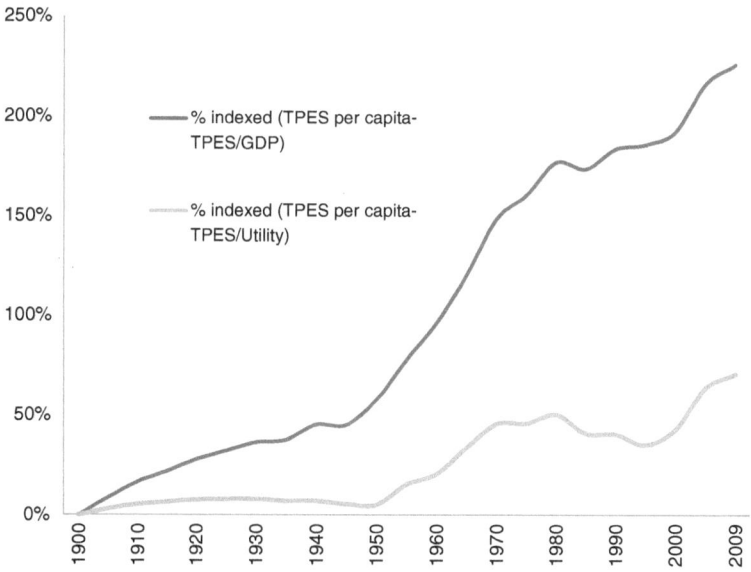

Fig. 6.2 Difference between DEC per capita and energy intensity

6.3   Energy Intensity of the USA for 1870–2005

6.3.1   The US Domestic Energy Consumption for 1870–2005

The US economy, among the most highly developed post-industrial economies, was the largest energy consumer and the biggest economy of the world until 2005.[2] Energy consumption in the USA has increased by 546.46 % during 1870–2005 (Table 6.2). Figure 6.3a depicts the trends in each major energy type consumed in the USA, measured in Peta-joule (PJ). Initially, from 1870, agricultural biomass and timber remain the most important energy resources, until 1906. After 1906, coal consumption increases more rapidly in comparison to biomass consumption. Oil and natural gas consumption remain relatively low through the same period. The period of the so-called Great Depression[3] (1929–1932) is characterized by significantly declining consumption trends in all energy resources. After the devastating impacts of the Great Depression and WWII, the year 1950 emerges as a milestone signaling a massive increase in energy consumption by the US economy. Indeed, the massive industrialization and the enormous building of infrastructure induced excessive energy consumption. Oil and natural gas become the predominant energy sources, while coal and biomass consumption increases as well after 1958. The oil shocks of 1973 and 1979 are clearly depicted in oil consumption trends; a first soft disturbance in oil consumption trends is observed in 1973, followed by a prolonged decline after the severe impact of the second oil crisis. The aftermaths of the second oil shock last until 1983, when a constant increase in oil consumption trends is initiated lasting until the end of the period being examined (2005). On the other hand, natural gas consumption follows a declining trend after 1972, with only a brief interruption during 1976–1985. Coal, electricity (from 1960), agricultural biomass, and timber consumption depict a general increasing trend, while fuel-wood consumption remains at relatively low levels compared with other energy inputs (Fig. 6.3a).

---

[2]However, after 2009, China takes the lion's share of total Primary Energy Consumption, and moreover, the Chinese economy has become the largest economy in the world, in terms of scale, in 2014 (Source: The Economist: http://www.economist.com/news/essays/21609649-china-becomes-again-worlds-largest-economy-it-wants-respect-it-enjoyed-centuries-past-it-does-not. Accessed June 2015).

[3]The economic crisis following the collapse of the Wall Street stock market in 1929 has been known since then as the Great Depression.

Table 6.2 Percentage change of the major indicators for indicative periods (USA)

	1900–2005 (%)	1900–1950 (%)	1950–1980 (%)	1980–2005 (%)
DEC/GDP	−77.58	−38.99	−35.09	−43.38
DEC/Utility	−13.24	21.61	−2.92	−26.51
DEC per capita	67.07	42.60	26.13	−7.11
DEC	546.46	184.25	88.63	20.57
GDP	2,783.13	365.89	190.58	112.97
GDP per capita	645.12	133.73	94.30	64.08
Population	286.93	99.33	49.55	29.80

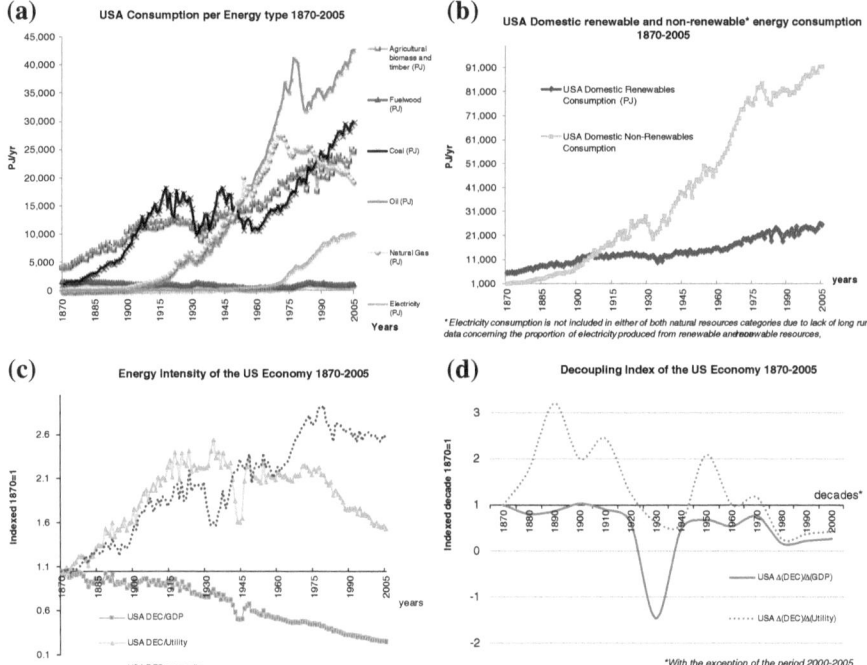

Fig. 6.3 The USA energy profile

6.3.2   USA Renewable and Non-renewable Energy Consumption for 1870–2005

The implications for the synthesis of energy resources induced by the transition of the US economy from the agrarian to a highly industrialized stage of development, is clearly depicted in the trends of the relative shares of renewable and non-renewable resources (Fig. 6.3b). We aggregate as US renewable energy use: agricultural biomass, timber, and fuel wood; and as US non-renewable energy consumption we aggregate: oil, natural gas, and coal including peat. Note that the renewables are underestimated to some extent, since data for electricity produced from renewable sources is not provided by the relevant database (Gierlinger and Krausmann 2012) and hence cannot be included in our estimates. The year 1906 is the milestone that signals the transition to the domination of non-renewable resources, ruled by fossil fuels consumption. The US economy is strongly reliant on non-renewable energy resources (91,609 PJ, in 2005—Fig. 6.3b), while renewable energy use remains low (25,736 PJ, in 2005—Fig. 6.3b).

6.3.3   The US Energy Intensity for 1870–2005

Figure 6.3c presents the estimates of the standard DEC/GDP indicator and the proposed DEC/Utility Energy Intensity (EI) indicators respectively, as well as the $DEC_{per\ capita}$ ratio (all indexed 1870 = 1) for the US economy. The DEC/GDP indicator presents a relative stability in 1880–1920; the WWII period is characterized by a strong EI decline lasting until the end of WWII in 1945; the short period of EI increment during 1946–1948 is followed by a constant and continuous decline of EI, until 2005. On the other hand, the proposed DEC/Utility indicator presents a quite dissimilar evolutionary path. Specifically, the proposed EI indicator shows a massive EI increase during 1870–1923; the "crisis period", 1932–1944 (post Great Depression and WWII), results in a strong EI reduction; the period 1945–1948 depicts again a strong EI increase, followed by relative stability in 1949–1977; after 1977, and until the end of the period under examination in 2005, the US DEC/Utility ratio presents a constant EI reduction.

Notably, DEC/Utility and $DEC_{per\ capita}$ ratios follow relatively similar patterns for the period 1870–1980, with the exception of the periods of the Great Depression and of WWII, during which the two ratios take completely different paths. During 1985–2005, the DEC/Utility ratio decreases markedly, whereas the $DEC_{per\ Capita}$ ratio shows a relative stability.

6.3.4   The US Decoupling Index, for 1870–2005

The US Decoupling Index (DI) is estimated for both the standard DEC/GDP and the proposed DEC/Utility indicators (Fig. 6.3d). Evidently, the $DI_{DEC/GDP}$ results in relative decoupling (values < 1) for the most of 1870–2005. Remarkably, an extreme value of absolute decoupling (DI = −1.45) can be detected in the 1930s and 1940s.

On the contrary, the $DI_{DEC/Utility}$ presents two major coupling periods (DI > 1) in 1870–1930 and 1950–1980, interrupted by a short relative decoupling period during 1930–1940. Strikingly, during 1980–2005, both $DI_{DEC/GDP}$ and $DI_{DEC/Utility}$ result in a similar relative decoupling pattern (Fig. 6.3d).

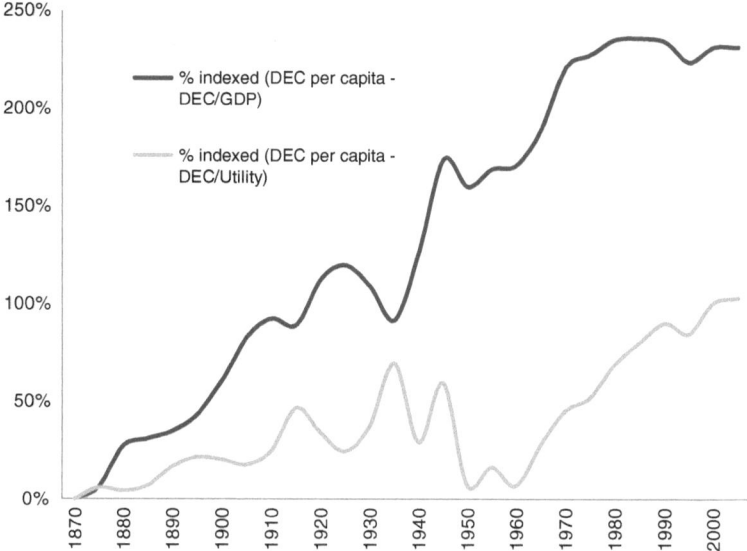

Fig. 6.4 Difference between DEC per capita and energy intensity

6.3.5   The Difference Between the Per Capita Energy Consumption and Energy Intensity Trends for the US Economy

Figure 6.4 presents the % change of the difference between the indexed values of DEC per capita and: (1) DEC/GDP and (2) DEC/Utility. The % difference of case (1) depicts an increasing trend throughout the period under examination (1870–2005), albeit with some short declining intervals during 1925–1935 and 1945–1950. On the other hand, case (2) depicts a period of fluctuating % change during 1870–1950; the period 1950–1960 reveals a remarkably marginal difference between DECper capita and DEC/Utility; finally, during 1965–2005, the difference between indexed DEC per capita and indexed DEC/Utility increases constantly.

6.4   Energy Intensity of Japan for 1878–2005

6.4.1   Japan's Domestic Energy Consumption for 1878–2005

Japan is a highly developed post-industrial economy based on knowledge and advanced know-how. What is more, Japan is considered to be an exceptional example of absolute decoupling between resource use and the economy over a prolonged period, as asserted by recent studies (Krausmann et al. 2011). Figure 6.5a presents the Japanese domestic energy consumption per energy type during 1878–2005. The DEC in Japan increased 2960.87 % during 1900–2005 (Table 6.3). Specifically, domestic consumption of oil increases dramatically during 1950–1973 (18,497.3 %), supporting the massive industrialization that took place after WWII. This trend is interrupted by the consequences of two oil shocks (1973 and 1979) which are clearly depicted in the reduction in oil consumption during 1973–1985; next, a short period of increase (1986–1992) is followed by a constant decline of oil consumption during 1993–2005. Natural gas and coal consumption present a constantly increasing trend from 1970s until 2005. Biomass use also increases constantly after WWII, to start decreasing smoothly from the mid-1980s on. Wood fuel use remains relatively stable for most of the 1878–1985, showing signs of small increments only during periods of turmoil (such as WWII and oil shocks), while after 1988, its use increases smoothly. Finally, electricity consumption increases constantly, except during the period 2000–2003 (Fig. 6.5a).

6.4.2   Japan's Renewable and Non-renewable Energy Consumption for 1878–2005

As a highly advanced post-industrial economy, Japan accomplished the transition from the agrarian stage to the industrial economy many decades ago. This transition required a dramatic increase in the use of non-renewable resources, the consumption of which progressively takes the lion's share after 1925, and takes off dramatically after 1950. The use of renewable resources increases constantly throughout the period examined, displaying a smooth decline only after 1995 (Fig. 6.5b).

6.4.3   Japan's Energy Intensity and Per Capita DEC Consumption for 1878–2005

Figure 6.5c represents the EI of the Japanese economy for the standard DEC/GDP and the proposed DEC/Utility ratios, both indexed to the base year of 1878. Evidently, there is a significant difference in the way these EI indicators evolve during 1878–2005. As indicated by Table 6.3, DEC/GDP decreased by −43.19 %, while DEC/Utility increased by 64.28 %, during 1900–2005. Specifically, the DEC/GDP indicator, with the exception of an increase during 1900–1914 and 1942–1945, demonstrates a decreasing trajectory, especially after 1945. Only recently (1985–2005) does the DEC/GDP ratio present a relative stability. On the contrary, the proposed DEC/Utility indicator follows an increasing trajectory throughout 1878–1973, only briefly interrupted by a period of relative stability

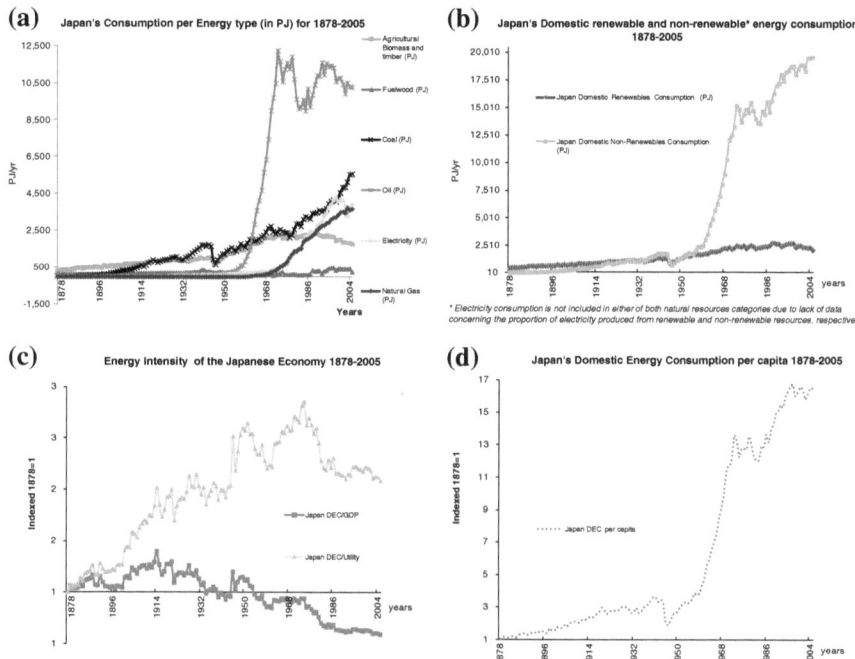

Fig. 6.5 Japan's energy profile

Table 6.3 Percentage change of the major indicators for indicative periods (Japan)

	1900–2005 (%)	1900–1950 (%)	1950–1973 (%)	1973–2005 (%)
DEC/GDP	−43.19	6.61	−15.64	−36.83
DEC/Utility	64.28	102.58	9.43	−25.89
DEC per capita	958.46	73.60	402.18	21.41
DEC	2,960.87	229.88	551.40	42.44
GDP	5,287.93	209.43	672.17	125.50
GDP per capita	1,763.17	62.84	495.29	92.21
Population	189.18	90.02	29.71	17.32

(1930–1943), and a declining period (1950–1958). The period 1974–1982 signals a strong decoupling trend, followed by relatively stability during 1987–2005.

The Japanese DEC $_{per\ capita}$ increased 958.46 % during 1900–2005 (Table 6.3). Interestingly, though, the per capita DEC of Japan shows a dramatic decrease during 1941–1945 (WWII), reflecting the devastating consequences of the WWII on the Japanese economy (Fig. 6.5d). However, the postwar period is characterized by a dramatically increasing DEC $_{per\ capita}$, interrupted only during 1973–1981 (the two oil crises), which increases again until 1996. Since 1997 and until 2005, DEC $_{per\ capita}$ presents a fluctuating stability (Fig. 6.5d).

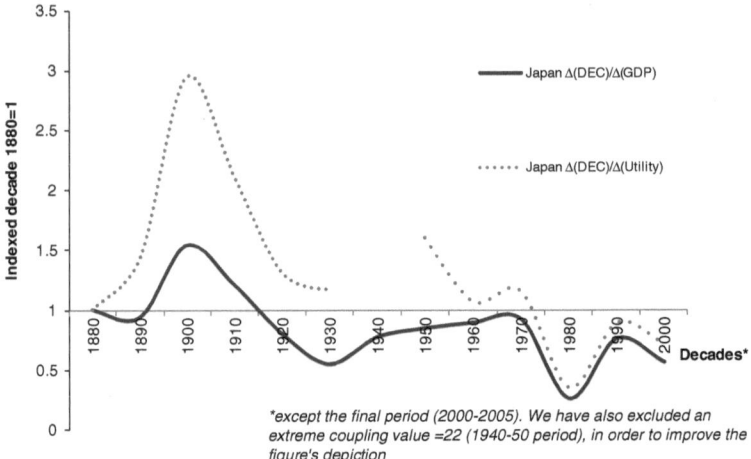

Fig. 6.6 Decoupling Index of the Japanese Economy 1880–2005

6.4.4 The Japanese Decoupling Index (DI), for 1870–2005

Figure 6.6 presents the Decoupling Index (DI) for both the standard (DEC/GDP) and the proposed (DEC/Utility) EI indicators. The DI(DEC/GDP) indicator results in only one coupling period, 1890–1920, while after 1920 and until 2005 it evolves within the "relative decoupling" area (1 > DI > 0). An exception could be detected during 1960–1970, where DI values are extremely close to 1, which is defined as the "border line" between relative decoupling and coupling. On the contrary, the DI (DEC/Utility) offers evidence of a long lasting coupling relationship during 1880–1970. Notably, during the 1980s and until the 2000s, both the standard and the proposed DIs show similar trajectories supporting a relative decoupling. It is worth mentioning that neither of the DI ratios indicates any period of absolute decoupling.

6.4.5 The Difference Between the Per Capita Energy Consumption and Energy Intensity Trends for the Japanese Economy

Figure 6.7 presents the % change of the difference between the indexed values of DEC per capita and: (1) DEC/GDP and (2) DEC/Utility. Both cases (1) and (2) follow similar patterns: increasing trends of % difference during 1880–2005; two declining periods in the differences occurring during 1945–1950 and 1970–1985.

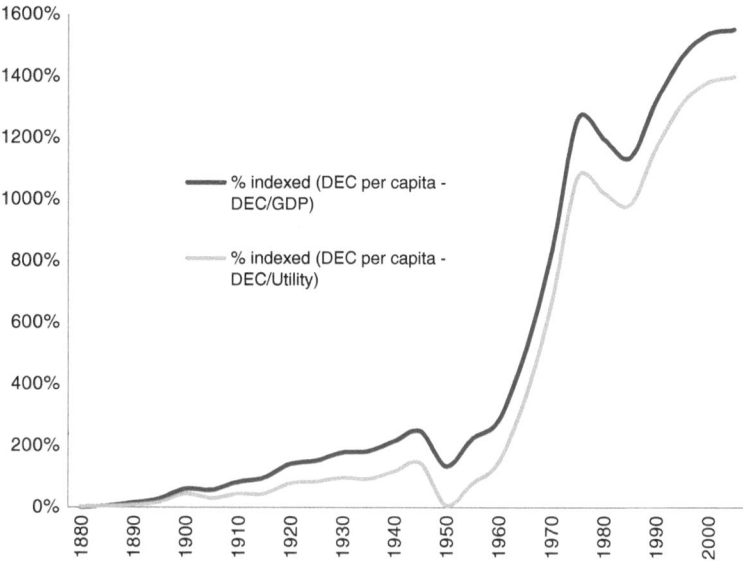

Fig. 6.7 Difference between DEC per capita and energy intensity

6.5 Extending the Estimates for the US and the Japanese Energy Intensity, Until 2013

Unveiling the essential characteristics of the energy intensity of the US and Japanese economies required the use of long run datasets covering more than a century of economic history for both countries. Unfortunately, these databases provide data only until 2005. In order to provide evidence from more up-to-date estimates, the present section utilizes the BP (2014) dataset which covers the period up to 2013, starting from 1965. We estimate the per capita DEC and the two EI indexes for the USA and Japan, for the period 1965–2013 (using 1965 = 1 as base year). The relevant trends are presented in Figs. 6.8 and 6.9 for the USA and Japan, respectively. Although differences in certain estimates, arising from the two data-sets, can be identified in the comparison between Figs. 6.3 and 6.8, as also between Figs. 6.5 and 6.9, in this section we emphasize the trends after 2005, based on the BP (2014) dataset.

The period after 2005 is characterized in the case of the USA by a decreasing EI for both indicators, strikingly more intensively for the DEC/GDP ratio. A quite interesting finding concerns the US DEC per capita index, which demonstrates a declining trend after 2005 (Fig. 6.8).

The Japanese DEC per capita demonstrates a decreasing trend around 2006, which is however reversed in 2009. Finally, after 2010, the DEC per capita decreases again, until 2013. After 2005 and until 2013, both of Japan's EI indices follow a decreasing trend which also characterizes the period before 2005 (Fig. 6.9).

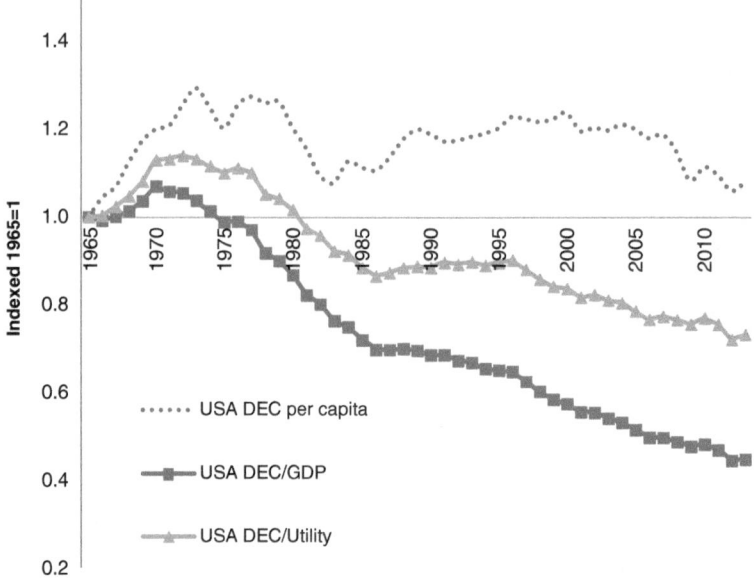

Fig. 6.8 USA energy intensity and DEC per capita for 1965–2013

6.6 The Energy Intensity of 10 Indicative Developed Countries from 1965 to 2013

6.6.1 Australia

Australia was the world's second largest coal exporter in 2011 and the third largest exporter of liquefied natural gas (LNG) in 2012 (EIA 2015).[4] Australia is heavily dependent on fossil fuels for its primary energy consumption, with coal representing the lion's share of its energy consumption since 1985, followed by oil of which the country is a net importer (BP 2014). Hydroelectricity and renewables consumption, although increasing after 2006, still remain at relatively low levels. Australia's per capita primary energy consumption decreases during 2005–2013, after a long period of extensive growth (Fig. 6.10—Annex I). For indicative numerical estimates, Fig. 6.10 and Annex I reveal substantial differences between the trajectories of DEC/GDP and the DEC/Utility EI indicators. While DEC/GDP decreases constantly after 1977, the proposed DEC/Utility indicator indicates a constant coupling trend lasting until the late 1990s, followed by a relative stability

---

[4]According to the EIA, Australia is among the most significant net hydrocarbon export countries in the OECD, exporting over 70 percent of its total energy production. (Source: http://www.eia.gov/countries/cab.cfm?fips=AS) (Accessed June 2015).

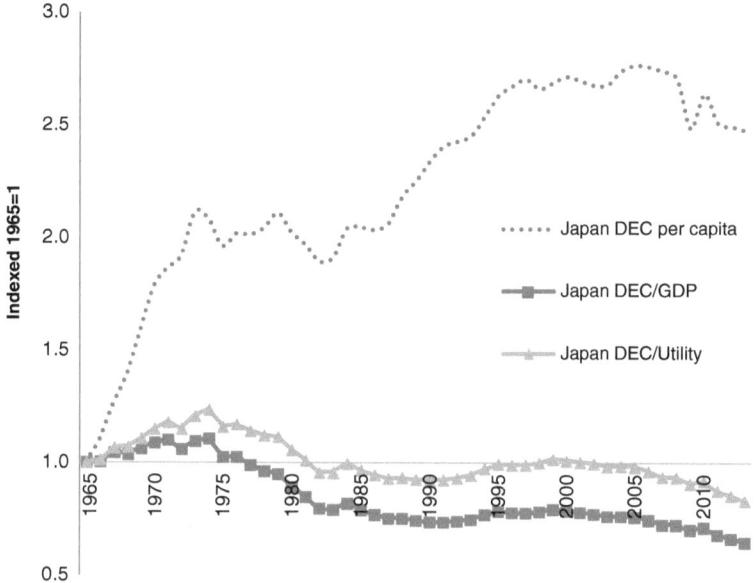

Fig. 6.9 Japan's energy intensity and DEC per capita for 1965–2013

until 2006. From 2006 and onwards, the DEC/Utility indicator initiates a declining trend similar to the trajectory of the per capita energy consumption in the same period.

6.6.2   Canada

Canada is considered to be among the most prosperous, wealthy and stable developed economies of the world (World Bank 2015). Furthermore, it is among the world's five largest energy producers. Canada controls the third-largest proven oil reserves in the world, after Saudi Arabia and Venezuela, and is the world's third-largest producer of dry natural gas. The country remains the principal source of the USA's electricity, oil and natural gas imports (EIA 2015).[5] Canada's domestic primary energy consumption is dominated by oil and natural gas consumption, followed by hydroelectricity production. It is worth mentioning that only China and Brazil produce more hydroelectricity than Canada (ibid).

Per capita energy consumption increases constantly until 2005 (Fig. 6.11 and Annex I); a sharp decline is observed during 2006–2009, followed again by increasing trends. Figure 6.11 exhibits the different ways in which the two EI

---

[5]Source: http://www.eia.gov/countries/cab.cfm?fips=CA (Accessed May 2015).

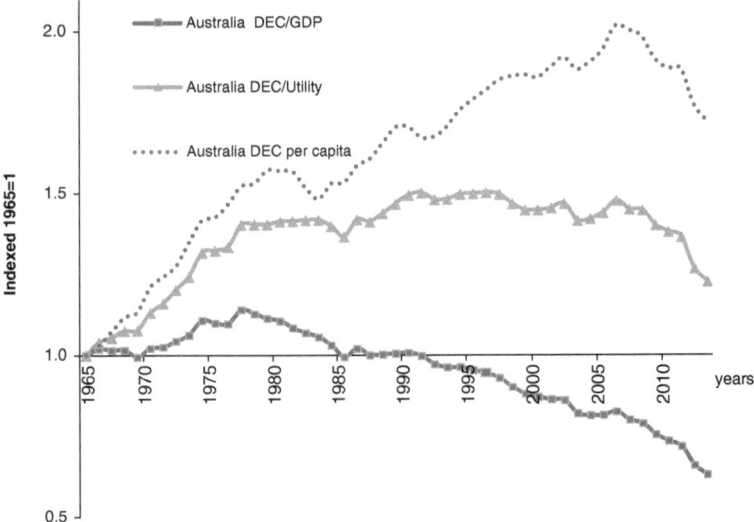

Fig. 6.10 Australia's energy intensity and DEC per capita energy consumption, for 1965–2013

indicators evolve. The DEC/GDP ratio follows a decoupling trend until 1990; smooth coupling during 1991–1996 and decreasing trends during 1997–2013. On the other hand, DEC/Utility indicates a coupling until 1996; decoupling during 1997–2001 and stability during 2002–2013.

6.6.3 Germany

Germany is among the five largest economies in the world and the largest national economy in Europe with a prominent role in international trade as it remains the third largest exporter in the world.[6] Germany is the largest energy consumer in Europe and the eighth-largest energy consumer in the world (EIA 2015).[7]

The German economy has presented remarkable improvement in energy efficiency as a result of technological advances and the implementation of energy efficiency regulations (ODYSSEE-MURE 2010).[8] The results are clearly reflected in the declining trends of per capita primary energy consumption in the late-1980s, as well as in the continuously declining EI trends after 1979, as evaluated by both

---

[6]Source Wikipedia: http://en.wikipedia.org/wiki/Economy_of_Germany (Accessed June 2015).

[7]Source EIA: http://www.eia.gov/countries/country-data.cfm?fips=GM (Accessed June 2015).

[8]Energy Efficiency Policies and Measures in Germany, ODYSSEE-MURE report, 2012. (Available at: http://www.isi.fraunhofer.de/isi-media/docs/x/de/publikationen/National-Report_Germany_November-2012.pdf) (Accessed June 2015).

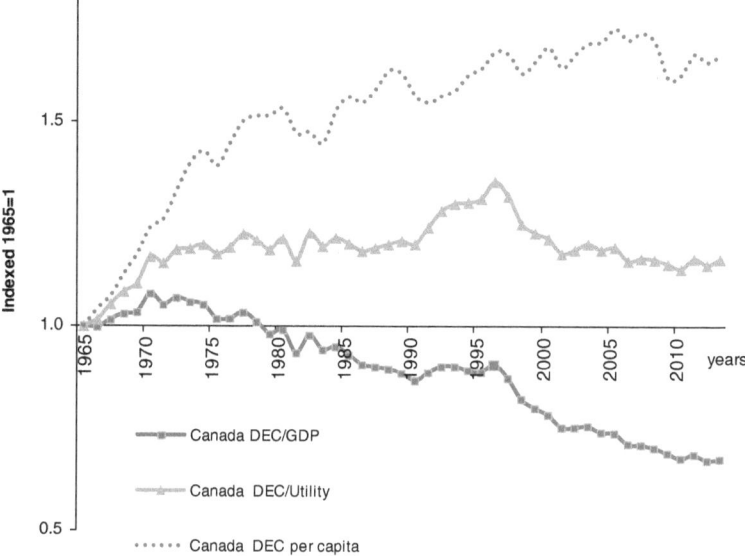

Fig. 6.11 Canada's energy intensity and per capita energy consumption, for 1965–2013

the DEC/GDP and the DEC/Utility indicators (Fig. 6.12 and Annex I). Remarkably, both the standard and the proposed EI indicators follow exactly the same evolutionary pattern throughout the whole period under examination.

6.6.4   Italy

Italy is one of the G8 countries and is the fourth largest economy in the European Union.[9] Italy is strongly dependent on fossil fuels imports, dominated by oil and natural gas consumption[10] as it is the second largest importer of natural gas in Europe, after Germany (EIA 2015). The per capita energy consumption in Italy increases constantly until 2004, with a short interval of decline during 1979–1983 (see Annex I for indicative numerical estimates). After 2005, the per capita energy consumption decreases until the end of the period under examination (2013) (Fig. 6.13). Remarkably, both the DEC/GDP and the DEC/Utility EI indicators display an almost identical declining trend during 1972–1985, that becomes less intensive during 1986–2013.

---

[9]Source Eurostat: http://ec.europa.eu/eurostat/statistics-explained/index.php/National_accounts_and_GDP (Accessed June 2015).

[10]Source EIA: http://www.eia.gov/countries/country-data.cfm?fips=IT&trk=m (Accessed June 2015).

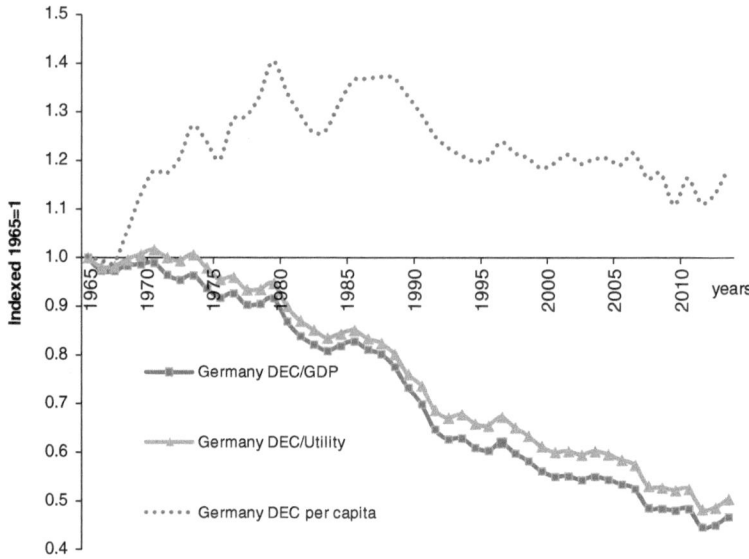

Fig. 6.12 Germany's energy intensity and per capita energy consumption, for 1965–2013

6.6.5 The Netherlands

The Netherlands is the sixth largest economy in the Eurozone.[11] The Netherlands is the second largest net exporter of natural gas in Europe after Norway; however the country remains a net importer of oil and coal (EIA 2015). Recent efforts of the Dutch government to promote energy efficiency and to develop renewable energy resources more intensively are reflected in the "Energy Agreement for Sustainable Growth" initiated in 2013 (OECD and IEA 2014). The per capita energy consumption in the Netherlands, with the exception of a sharp decline during 1979–1983, increases until 2005, while the recent period of 2010–2013 demonstrates a sharp decline (Fig. 6.14). The EI trends for both DEC/GDP and DEC/Utility present a smooth declining trend, after 1973, with short intervals of relative stability and hints of periodic smooth coupling (Fig..6.14 and Annex I for indicative numerical results).

[11]Source Eurostat: http://ec.europa.eu/eurostat/statistics-explained/index.php/National_accounts_and_GDP (Accessed June 2015).

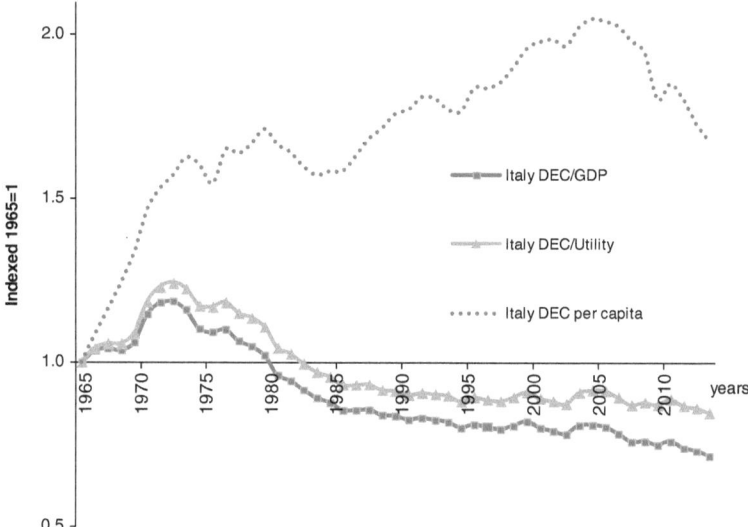

Fig. 6.13 Italy's energy intensity and per capita energy consumption, for 1965–2013

6.6.6 Norway

Norway is considered to be among the most prosperous economies of the world.[12] Furthermore, Norway is the largest oil producer in Europe and the world's third largest natural gas net-exporter (EIA 2015).[13] Rich in natural endowments as it is, Norway saves revenues from the petroleum sector in the world's largest sovereign wealth fund, the so-called Government Pension Fund Global (GPFG), valued at over $830 billion in January 2014.[14] The main aim of GPFG is to deliver the highest possible long-term returns, arising from the invested revenues of the petroleum sector, in order to build wealth for the future generations of Norway.[15] Per capita energy consumption in Norway follows a constantly increasing trend until 2000, resulting in periodically fluctuating trajectories in recent years, during 2000–2013. On the other hand, both the EI indicators follow similar trajectories indicating relative stability until 1990, when a period of relative decoupling begins (Fig. 6.15 and Annex I).

---

[12]Source: http://www.heritage.org/index/country/norway (Accessed June 2015).

[13]Source EIA: http://www.eia.gov/beta/international/analysis.cfm?iso=NOR (Accessed June 2015).

[14]Source http://www.indexmundi.com/norway/economy_profile.html (Accessed May 2015).

[15]Source: http://www.nbim.no/globalassets/reports/2013/annual-report/annual-report_2013_web.pdf (Accessed June 2015).

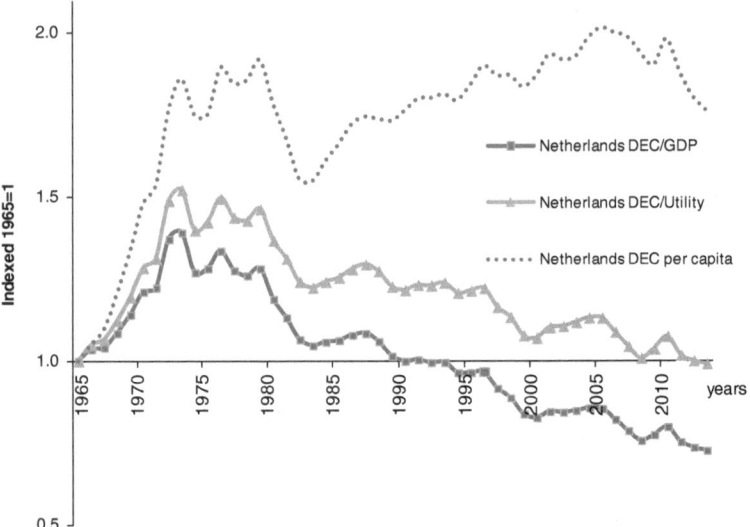

Fig. 6.14 The Netherlands' energy intensity and per capita energy consumption, for 1965–2013

6.6.7   South Korea

South Korea has demonstrated impressive growth during the last four decades, inducing a transition from a poor developing country back in the 1960s into a high-tech developed economy[16,17] The South Korean economy, due to lack of domestic energy reserves, depends on imports for up to 97 % of its energy use, being the world's ninth largest energy consumer in 2011 (EIA 2015). The per capita energy consumption of South Korea increases intensively throughout the period under examination (Fig. 6.16b and Annex I). The EI trends for both the DEC/GDP and the DEC/Utility indicators depict a strong coupling during 1965–1997, a sharp decoupling period during 1998–2006, and finally a period of relative stability, during 2007–2013 (Fig. 6.16a and Annex I).

6.6.8   Spain

Spain is the fourth largest economy in the Eurozone and has recently returned to positive growth (in the second half of 2013) after a prolonged recession (OECD 2014). Spain is the fifth largest energy consumer in Europe and a net importer of

---

[16]Source: http://www.forbes.com/places/south-korea/ (Accessed, May 2015).
[17]Source: http://www.economist.com/news/finance-and-economics/21647323-once-fearsome-economy-struggles-fend-deflationary-funk-tiger-winter (Accessed, May 2015).

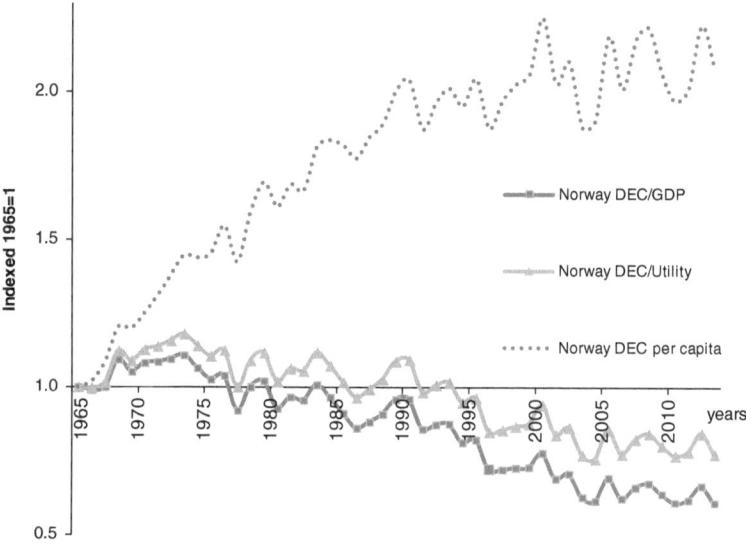

Fig. 6.15 Norway's energy intensity and per capita energy consumption, for 1965–2013

fossil energy resources (EIA 2015). The per capita energy consumption in Spain follows a constantly increasing path until 2007, with a sharp decline observed during 2008–2013 probably as a result of the financial crisis which deeply affected the Spanish economy (Fig. 6.17b and Annex I). The DEC/Utility indicator results in two strong coupling periods (1965–1980; 1991–2004), and two decoupling periods (1981–1990; 2005–2013). Similar trends are supported by the DEC/GDP indicator. The declining trend of both EI indicators, during 2005–2013, may be related to the global financial crisis of 2008 (Fig. 6.17a).

6.6.9   Taiwan

Taiwan is ranked as the 18th largest economy of the world, in terms of aggregate GDP (2012).[18] Taiwan is a net-importer of oil and coal (75 % of Taiwan's primary energy consumption in 2013), due to its limited domestic energy resources (EIA 2015). Per capita energy consumption is steadily increasing until 2007, followed by stability in recent years 2008–2013 (Fig. 6.18b and Annex I). The EI trends, estimated by the DEC/Utility indicator, result in a strong coupling period (1965–1980), a period of relative stability (1981–1997), followed again by intensive coupling

---

[18]Source: https://www.wto.org/english/res_e/publications_e/wtr12_e.htm (Accessed June 2015).

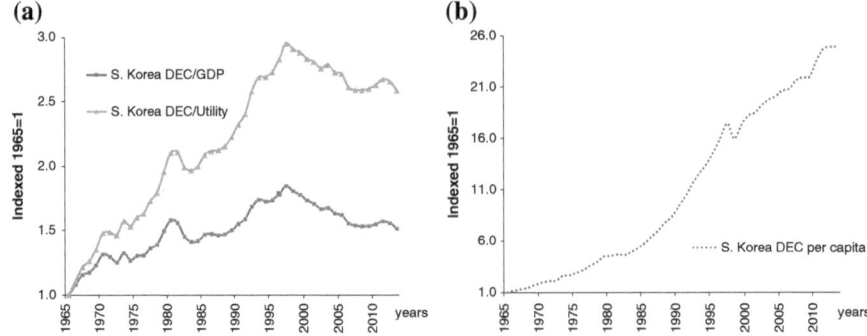

Fig. 6.16 South Korea's **a** energy intensity and; **b** per capita energy consumption, for 1965–2013

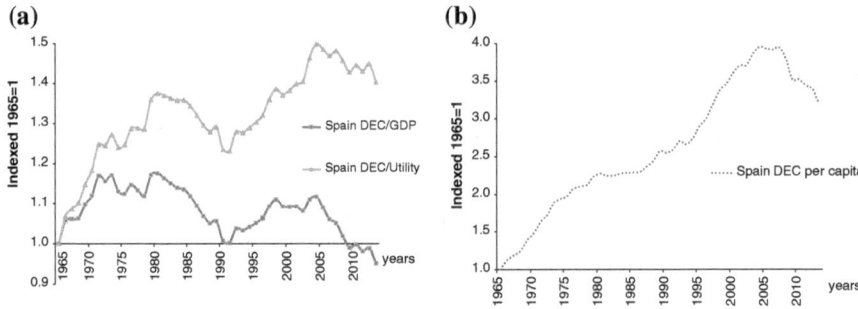

Fig. 6.17 Spain's **a** energy intensity and; **b** per capita energy consumption, for 1965–2013

(1998–2003), to end up with a strong decoupling trend (2005–2013). On the other hand, the DEC/GDP ratio depicts a smooth coupling trend (1965–1980), a gradual decoupling trend (1981–1997) briefly interrupted by a coupling period (1998–2003), and ending with an intensive decoupling period during 2004–2013 (Fig. 6.18a, Annex I).

6.6.10 The United Kingdom

The United Kingdom (UK) is the second largest economy in the European Union in terms of GDP (2014).[19] Furthermore, the UK is the largest oil producer and the second largest natural gas producer in the European Union; nevertheless, after many years of being an oil and natural gas net exporter, the UK became, after 2004, a

---

[19]Source: http://ec.europa.eu/eurostat/statistics-explained/index.php/National_accounts_and_GDP (Accessed June 2015).

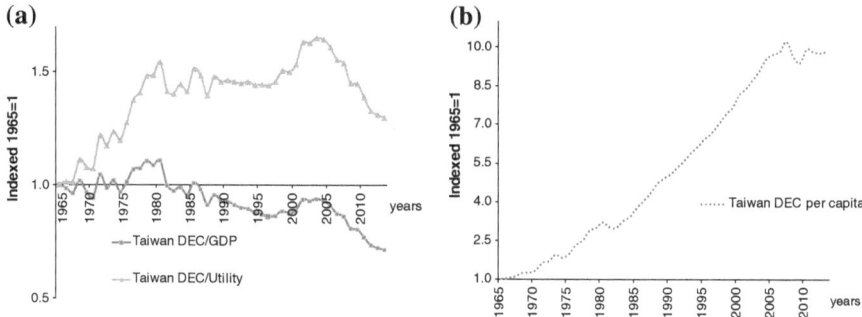

Fig. 6.18 Taiwan's **a** energy intensity and; **b** per capita energy consumption, for 1965–2013

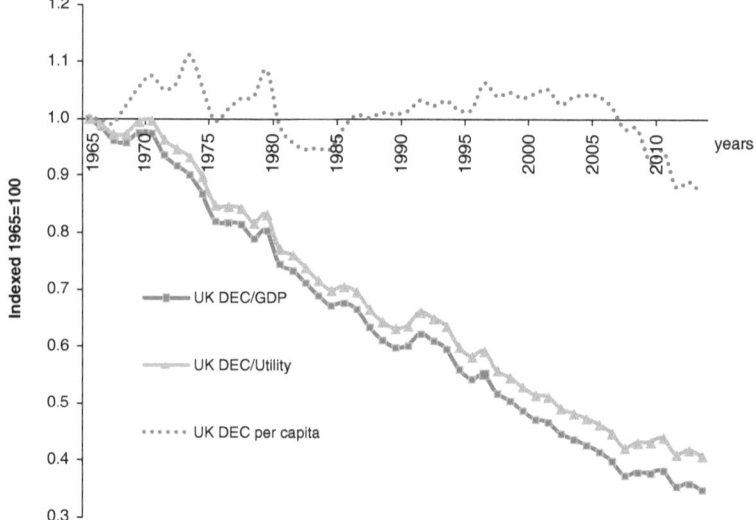

Fig. 6.19 The United Kingdom's energy intensity and per capita energy consumption, for 1965–2013

net-importer for both fuels due to the peak of domestic oil and natural gas consumption that occurred in late 1990s (EIA 2015).[20] The UK's per capita energy consumption presents fluctuating trends until 1980; nevertheless, a relative stability is displayed during 1980–2005, ending up with a sharp decline after 2006 (Fig. 6.19, Annex I). The UK's EI, as estimated by the standard and the proposed framework, result in remarkably similar evolutionary patterns for both EI indicators. The EI displays a constantly declining trend, with only three brief periods of relative stability: 1975–1979; 1989–1991; 2007–2013 (Fig. 6.19 and Annex I for indicative numerical estimates).

[20]Source: EIA (http://www.eia.gov/countries/country-data.cfm?fips=uk) (Accessed April 2015).

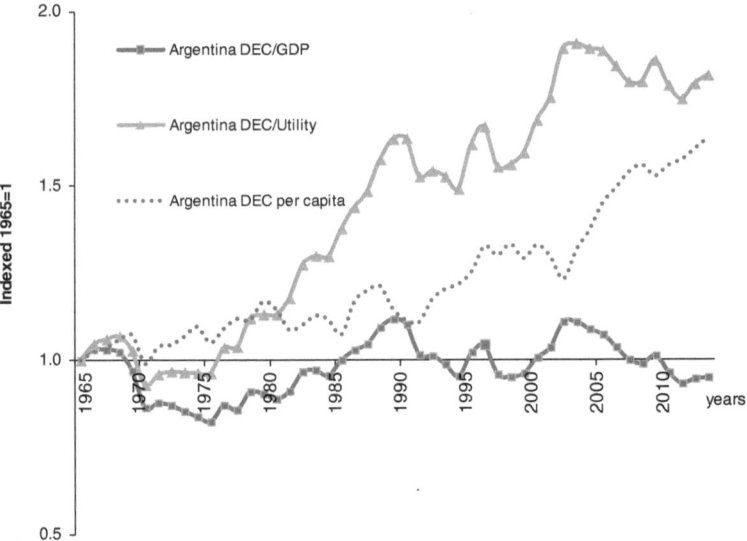

Fig. 6.20 Argentina's energy intensity and per capita energy consumption, for 1965–2013

6.7   The Energy Intensity of 10 Indicative Developing Countries for 1965–2013

6.7.1   Argentina

Argentina is the third largest economy in Latin America, in terms of GDP (World Bank 2015).[21] Despite the impacts of the extended economic crisis during 1998–2002, the Argentinean economy has been growing steadily during the last decade (ibid). Argentina was the largest dry gas producer in Latin America in 2013 (EIA 2015). After a period of fluctuating trends during 1965–2002, the per capita energy consumption in Argentina increases almost steadily until 2013, with the exception of a short interval during 2008–2009 which probably reflects the consequences of the economic recession at that time (Fig. 6.20). The DEC/GDP ratio presents two main coupling periods during 1975–1990 and 1998–2002, and a prolonged decoupling in recent years 2003–2013. On the other hand, the DEC/Utility ratio displays a constant coupling trend during 1970–1990, fluctuations during 1991–1998, a strong coupling in 1999–2003, and fluctuations again in the recent years 2004–2013 (Fig. 6.20 and Annex I). Interestingly, the DEC per capita presents similar trajectories to the DEC/Utility ratio.

---

[21]Source: http://www.worldbank.org/en/country/argentina/overview (Accessed June 2015).

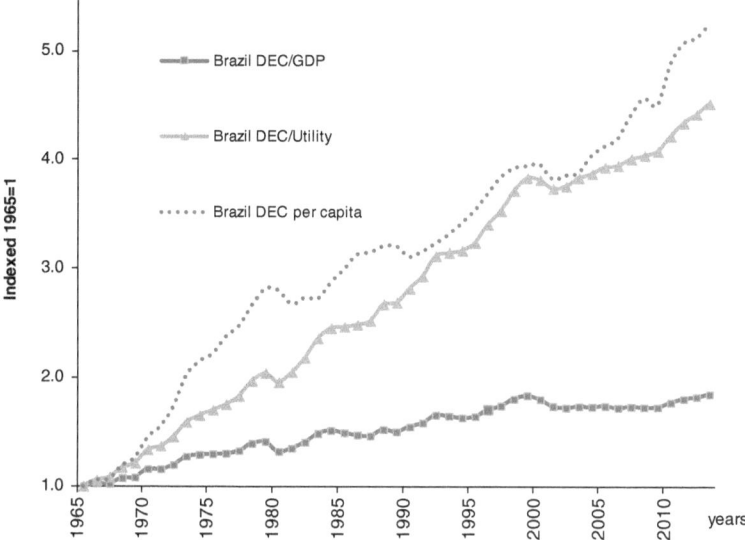

Fig. 6.21 Brazil's energy intensity and per capita energy consumption, for 1965–2013

6.7.2 Brazil

Brazil is the world's seventh largest economy (in terms of GDP), and the largest country (in terms of area and population) in Latin America (World Bank 2015). In 2013, Brazil was the eighth largest energy consumer and the tenth largest energy producer in the world (EIA 2015). Oil consumption and hydroelectricity are by far the most predominant energy flows feeding the economic growth of Brazil (ibid). Figure 6.21 presents the per capita energy consumption in Brazil and the EI trends for the DEC/GDP and DEC/Utility indicators. The DEC per capita ratio increases almost constantly throughout the period under examination 1965–2013. Similarly, both EI indicators result in constantly increasing energy intensity until 2000. After 2000, the DEC/GDP indicator depicts a period of relative stability during 2001–2010, followed by a smooth increase during 2011–2013. On the contrary, after a short period of slowdown observed in 2000–2002, the DEC/Utility indicator displays a steep coupling in 2003–2013. Notably, the DEC/Utility indicator presents a remarkably similar evolutionary pattern to the per capita DEC throughout the period under examination (Fig. 6.21, Annex I).

6.7.3 China

China is the most rapidly growing economy (According to the journal *The Economist*, the Chinese economy became, in 2014, the largest economy in the

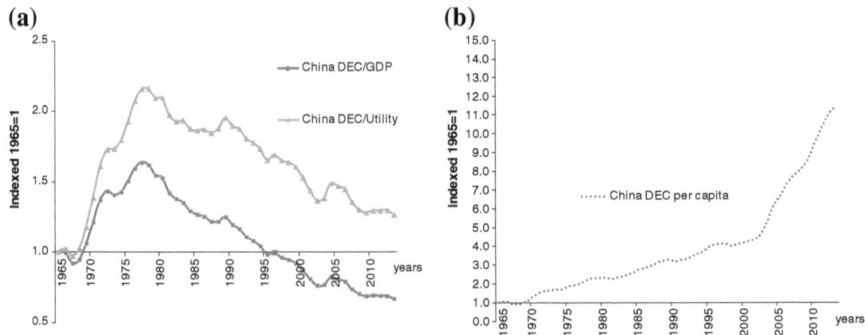

Fig. 6.22 **a** China's energy intensity; and **b** China's per capita energy consumption, for 1965–2013

world in terms of scale[22]) as well as the most populous country in the world (World Bank 2015). After 2010, China became the largest energy consumer in the world, with coal being the predominant energy resource that feeds the growth of the Chinese economy (EIA 2015). Per capita energy consumption in China presents a dramatic and accelerating increase after 2001 (Fig. 6.22b, and Annex I). On the contrary, both EI indicators reveal a steep coupling trend until 1978. During 1979–2002, a gradual decoupling in EI trends is depicted by both indicators which evolve in stark contrast with the DEC per capita trends. Finally, the period 2008–2013 reveals stabilization in the EI trends for both EI indicators (Fig. 6.22a, Annex I). Overall, the estimates of both EI indicators and energy metabolism (DEC per capita) indicate the particular characteristics of the Chinese socio-economic system which requires an in-depth analysis, something which is beyond the scope of the present volume.

6.7.4   India

India is the fourth largest economy and the second most heavily populated country of the world (World Bank 2015).[23] India is moving rapidly towards a transition from the agrarian stage of growth to massive industrialization. According to the World Bank (2015) and Singh et al. (2012), massive investments will be needed to facilitate the infrastructure and housing necessities. Undoubtedly, the future of India's transition towards the path of development requires increasing energy and material inputs (ibid). India was the fourth largest energy consumer in the world in

---

[22]Source: The Economist: http://www.economist.com/news/essays/21609649-china-becomes-again-worlds-largest-economy-it-wants-respect-it-enjoyed-centuries-past-it-does-not (Accessed June 2015).

[23]Source: http://www.worldbank.org/en/country/india/overview (Accessed June 2015).

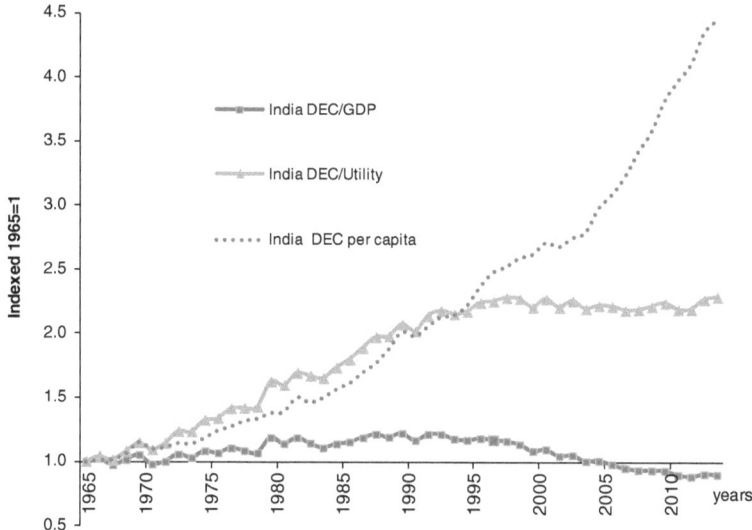

Fig. 6.23 India's energy intensity and per capita energy consumption, for 1965–2013

2011 (EIA 2015). The sharp increase of India's per capita energy consumption, especially after 2001, is clearly pictured in Fig. 6.23. Remarkably, the DEC/Utility indicator follows a coupling trend similar to the DEC per capita ratio until the mid-1990s, while during 1995–2013 there is relative stability with a hint of a new increase during 2009–2013. The DEC/GDP ratio, although initially depicts a smooth coupling trend, ends up in an almost constant decoupling trend during 1992–2010 (Fig. 6.23, Annex I).

6.7.5   Indonesia

Indonesia is the tenth largest economy of the world, in terms of GDP, and the fourth most populous country of the world (World Bank 2015). Indonesia has been a net importer of oil since 2004; however the country remains a net exporter of coal and natural gas (EIA 2015). Per capita energy consumption in Indonesia has been constantly increasing since 1971, with however a period of slow-down occurring during 2003–2008. DEC/Utility depicts a strong coupling trend for the period 1965–2003, followed by a period of fluctuating stability during 2004–2013. The DEC/GDP depicts a smooth coupling trend for the period 1965–2003, followed by a period of a smooth decoupling, during 2004–2013 (Fig. 6.24). As a result, the two EI indexes follow similar trajectories with intensifying coupling until the mid-2000s, with the more recent years characterized by relatively increasing energy efficiency (Fig. 6.24, Annex I).

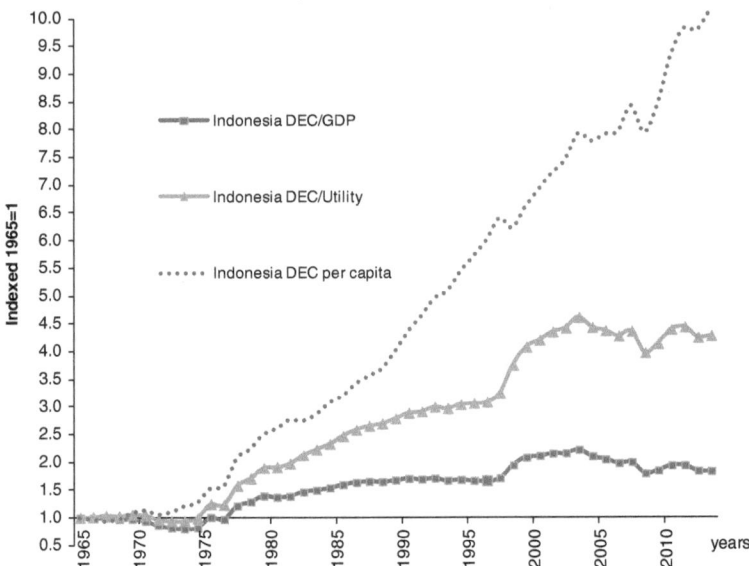

Fig. 6.24 Indonesia's energy intensity and per capita energy consumption, for 1965–2013

6.7.6   Malaysia

Malaysia, which is the 28th largest economy of the world, is considered an exceptional example of an economy with remarkable growth rates, at an average 7 % per year for more than 25 years (World Bank 2015).[24] This growth was accompanied by a dramatic reduction in poverty (ibid). Furthermore, Malaysia is the world's second largest exporter of LNG and the second largest oil and natural gas producer in South Asia (EIA 2015). The energy industry plays a decisive role in the Malaysian economy, representing 20 % of total GDP in 2014 (EIA 2015), while the country remains a leading exporter of electronic parts, appliances, and other knowledge-based devices (World Bank 2015). The per capita energy consumption in Malaysia increases constantly and intensively during 1965–2013 (Fig. 6.25, Annex I). Similarly, the DEC/Utility ratio depicts a constant coupling trend, with a period of relative stability during 2005–2013. On the other hand, the DEC/GDP presents a period of relative stability and smoothly increasing trends during 1986–2003, while since 2003 a smooth declining trend has been occurring (Fig. 6.25, Annex I).

---

[24]Source: http://www.worldbank.org/en/country/malaysia/overview (Accessed May 2015).

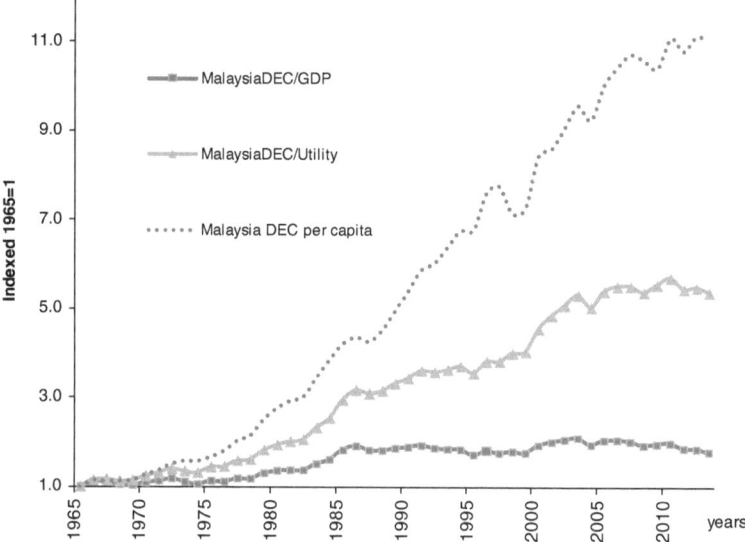

Fig. 6.25 Malasia's energy intensity and per capita energy consumption, for 1965–2013

6.7.7   Mexico

Mexico is the largest economy in Latin America and among the most important emerging economies (World Bank 2015).[25] According to EIA (2015), Mexico is one of the ten largest oil producers in the world, the third largest in the Western Hemisphere.[26] However, the contribution of oil is constantly decreasing in the total energy mix of the country, being gradually replaced by natural gas, a resource of which Mexico is a net importer (EIA 2015). The per capita energy consumption in Mexico is constantly increasing throughout the period under examination, however with some fluctuations after 1995 and an indication of stability after 2010 (Fig. 6.26). The DEC/Utility indicator displays an increasing coupling trend, sharing substantial similarities with the per capita energy consumption trends. On the other hand, the DEC/GDP ratio presents a long lasting period of relative stability in EI trends, during 1987–2013 (Fig. 6.26—see also Annex I for percentage changes). As a result, the two EI indicators follow substantially different evolutionary patterns reflecting different estimates for the EI of Mexico's economy.

---

[25]Source: http://www.worldbank.org/en/country/mexico (Accessed May 2015).

[26]Source: http://www.eia.gov/countries/cab.cfm?fips=MX (Accessed May 2015).

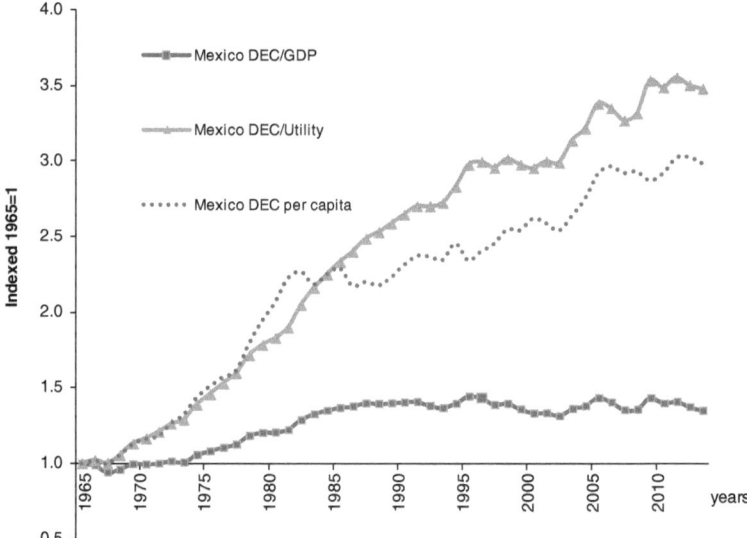

Fig. 6.26 Mexico's energy intensity and per capita energy consumption, for 1965–2013

6.7.8   Pakistan

The main characteristics of Pakistan's economy are moderate economic growth trends, weak performance of the manufacturing sector and stagnant exports (Asian Development Bank 2015). Nevertheless, the World Bank projects a growth rate of 6 % in 2015 (World Bank 2015).[27] Energy shortages and insecurity are a structural characteristic of Pakistan (EIA 2015). The country is a net importer of oil and coal, while it consumes domestically most of the natural gas it produces (ibid). Remarkably, the DEC per capita and the DEC/Utility ratios follow identical evolutionary paths showing, after 1972, a continuously and intensively increasing trend, except for the period 2005–2013 where a short smooth decline is observed. In contrast, the DEC/GDP indicator presents a smoothly increasing trend until 2004, while it gradually decreases during 2005–2013 (Fig. 6.27, Annex I).

6.7.9   Thailand

Thailand has been one of the most widely cited development success stories, with sustained growth and impressive poverty reduction, particularly in the 1980s

---

[27]Source: http://data.worldbank.org/country/pakistan (Accessed June 2015).

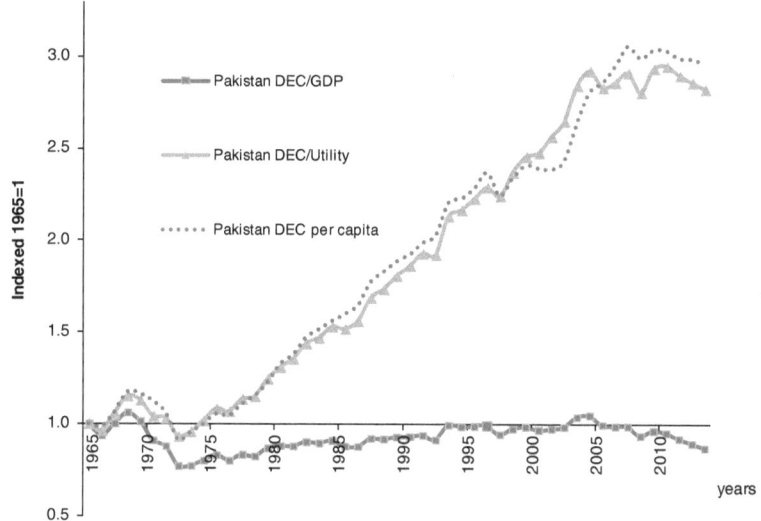

Fig. 6.27 Pakistan's energy intensity and per capita energy consumption, for 1965–2013

(World Bank 2015; Asian Development Bank 2015). Although the country is a natural gas and oil producer, it remains a net importer of fossil fuels in order to meet its increasing demand for energy. Thailand's DEC per capita ratio increases intensively throughout the period under examination, with only a short interruption occurring during 1997–1998 (Fig. 6.28b, Annex I). Similarly to the DEC per capita trends, the DEC/Utility indicator displays strong coupling, especially after 1982. On the other hand, the DEC/GDP ratio depicts a more moderate coupling compared to the DEC/Utility during 1982–1997, while the period 1998–2013 results in relative stability with a suggestion of a smooth increase (Fig. 6.28a, Annex I).

6.7.10   Turkey

Turkey was the 18th largest economy of the world in 2014. Although it is a member of OECD and G20, the country is still defined as a developing one (World Bank 2015).[28] Turkey has enjoyed an increasing growth rate for many years, and it recovered quickly after the financial crisis of 2008. Turkey's energy demand has increased strongly during the last few years, a trend which is expected to increase more intensely in the future, underlining the characteristics of the country as both a large energy consumer and an energy transit hub. Indeed, Turkey plays a key role in oil and natural gas transportation from Russia, the Caspian Sea, and the Middle East

---

[28]Source: http://www.worldbank.org/en/country/turkey/overview (Accessed June 2015).

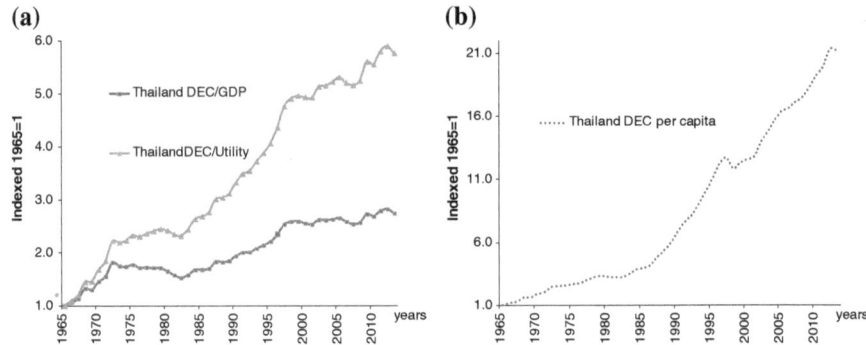

Fig. 6.28 Thailand's **a** energy intensity and; **b** per capita energy consumption, for 1965–2013

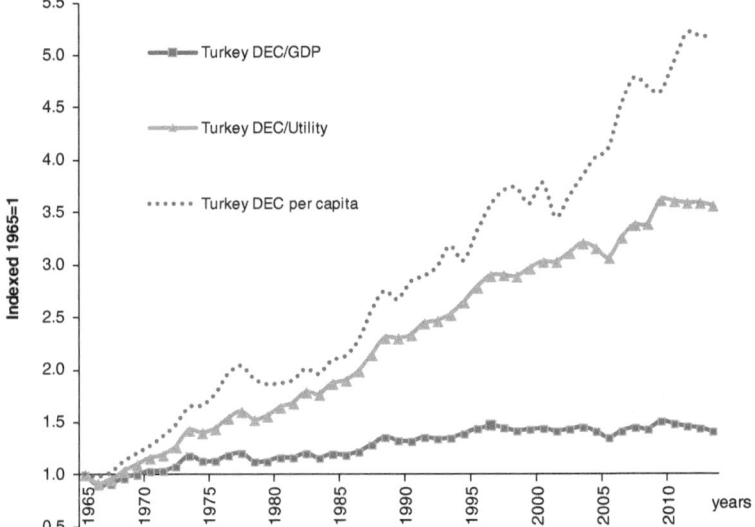

Fig. 6.29 Turkey's energy intensity and per capita energy consumption, for 1965–2013

to Europe, since it is becoming an important route of pipelines for oil and natural gas (EIA 2015).[29] The DEC per capita increases constantly in Turkey until 2013, with a short interruption during 1998–2001. The DEC/Utility indicator performs a constant coupling trend which presents substantial similarities to the evolutionary pattern of the per capita DEC ratio. Notably, both DEC per capita and DEC/Utility show a hint of decrease in 2010–2013. In contrast, the DEC/GDP indicator presents a far more moderately increasing trend than the DEC/Utility ratio over the whole period examined (Fig. 6.29, Annex I).

---

[29]Source: http://www.eia.gov/beta/international/analysis.cfm?iso=TUR (Accessed June 2015).

References

Asian Development Bank (ADB). (2015). *Asian development outlook 2015*. Financing Asia's future growth. Philippines: ADB. Available at http://www.adb.org/sites/default/files/publication/154508/ado-2015.pdf. Accessed June 2015.

B.P. (2014). *Statistical review of world energy 2014*. Available at http://www.bp.com/en/global/corporate/about-bp/energy-economics/statistical-review-of-world-energy.html. Accessed in February 2015.

Energy Information Administration (EIA). (2015). *International energy data and analysis*. All countries' overviews are available on-line at http://www.eia.gov/beta/international/?fips=as. Accessed between May and August 2015.

Geary, R. C. (1958). A note on comparisons of exchange rates and purchasing power between countries. *Journal of the Royal Statistical Society, 121*(1), 97–99.

Gierlinger, S., & Krausmann, F. (2012). The physical economy of the United States of America. *Journal of Industrial Ecology, 16*(3), 365–377.

Khamis, S. H. (1972). A new system of index numbers for national and international purposes. *Journal of the Royal Statistical Society, 135*(1), 96–121.

Krausmann, F., Gingrich, S., Eisenmenger, N., Erb, K. H., Haberl, H., & Fischer-Kowalski, M. (2009). Growth in global materials use, GDP and population during the 20th century. *Ecological Economics, 68*(10), 2696–2705.

Krausmann, F., Gingrich, S., & Nourbakhch-Sabet, R. (2011). The metabolic transition in Japan. *Journal of Industrial Ecology, 15*(6), 877–892.

Maddison, A. (2008). *Statistics on world population, GDP and per capita GDP, 1–2008 AD*. Retrieved June, 2011, from Angus Maddison (1926–2010): http://www.ggdc.net/MADDISON/oriindex.htm.Accessed in June 2011.

ODYSSEE-MURE. (2010). Monitoring of EU and national energy efficiency targets. Energy Efficiency Policies and Measures in Germany in 2012. Report is available at http://www.isi.fraunhofer.de/isi-wAssets/docs/x/de/publikationen/National-Report_Germany_November-20-12.pdf. Accessed May 2015.

OECD. (2014). Economic surveys: Spain overview. Available at http://www.oecd.org/Spain. Accessed June 2015.

OECD and IEA. (2014). *World energy outlook 2014*. Paris: International Energy Agency. Executive summary available at https://www.iea.org/publications/freepublications/publication/WEO_2014_ES_English_WEB.pdf. Accessed January 2015.

Singh, S. J., Krausmann, F., Gingrich, S., Haberl, H., Erb, K. H., Lanz, P., et al. (2012). India's biophysical economy, 1961–2008. Sustainability in a national and global context. *Ecological Economics, 76*, 60–69.

The Conference Board Total Economy Database. (2014). Available at http://www.conference-board.org/data/economydatabase/.

World Bank. (2015). *Countries overview*. Available on-line at http://www.worldbank.org/en/country. Accessed between May and August 2015.

Chapter 7
Discussion and Conclusions

> *Once upon a time, throughout the heyday of classical*
> *economics, demography belonged to political economy. The*
> *supply of labor was one of the most important endogenous*
> *variables in the systems of Smith, Malthus, Mill and Marx. One*
> *feature of neoclassical economics that distinguishes it from the*
> *classical version is the removal of population as a variable*
> *subject to economists' analysis.*
>
> Paul Samuelson, 1985

Abstract The fundamentally different trajectories of the Energy Intensities, as estimated by the proposed and the prevailing framework, are discussed and explanations are provided. Grouping economies according to their Energy Intensities may provide a basis for delineating those characteristics that may induce an actual decoupling and lessen the dependency of economy on energy. Challenges requiring further analysis are identified.

Keywords Energy intensity · Energy dependency · Growth-degrowth dialogue

This chapter discusses the empirical estimates of Energy Intensity (EI) trends which were presented systematically in Chap. 6, and attempts to classify economies according to their energy requirements and then to draw general conclusions concerning the link between energy and growth. In this book, the focus of the analysis lies in the macro trends of EI which indicate the essential dependency of an economy on energy inputs. Short run, occasional shifts lie outside the interest of the current analysis, in so far as they do not imply structural changes in the link between Energy and the Economy. The core estimates of EI are those based on the proposed Energy (E)/Utility indicator while additional estimates offer a basis for further investigation of the energy dependency of the economies.

7.1 Comparing E/GDP and E/Utility Estimates

An interesting classification of economies could be based on the comparison between the trends in the E/GDP and E/Utility indexes. Clearly, a difference between the values of the two ratios is to be expected as a result of their algebraic

© The Author(s) 2016
K. Bithas and P. Kalimeris, *Revisiting the Energy-Development Link*,
SpringerBriefs in Economics, DOI 10.1007/978-3-319-20732-2_7

structure. The real focus of the analysis should be on the difference between the trajectories that the indexed values of each ratio follow. This difference is substantial and important in most of the economies studied.

The three economies with long run data present substantial differences in the trends of the two EI indicators. For the global economy, the E/GDP ratio follows a constantly declining path which intensifies after 1950. In contrast, the E/Utility ratio grows intensively until 2000 at which point stabilization is achieved. It is noteworthy that the trends in E/Utility are very similar to those of the per capita use of energy at the global level throughout the 20th century. However, global per capita use still increases, albeit at a lower rate, after 2000 when E/Utility stabilizes.

The giant economies of the USA and Japan have also different trajectories in E/GDP and E/Utility until the mid-1970s. The mid-1970s are a milestone as during this period the E/Utility ratio begins to demonstrate declining trends which approximate those of the standard E/GDP index. In the case of the USA, this trajectory is supported by a declining 'Energy per capita' index. Remarkably, per capita Energy use in Japan follows an almost constantly increasing path, interrupted briefly from 1973–1983. In Japan, this increasing trend persists, even when both EI indices decline or remain relatively stable.

For the countries with short run data, similar trajectories of E/GDP and E/Utility are identified for Brazil, China, Indonesia, Thailand, Norway, the Netherlands, Italy, South Korea, Germany, Spain and the UK. Of these countries, only Brazil presents increasing Energy Intensity, for both indicators, over almost the entire period under examination. In all the other economies, the EI follows a declining trend after the mid-1970s with the exception of Spain and South Korea which initiated their decoupling trend around 2000, and Indonesia that stabilized its EI in the same year. For the group of the economies with similar trends in E/GDP and E/Utility, the energy per capita index shares substantial similarities with the EI trends. Only rapidly developing China and South Korea demonstrate a constantly intensifying energy use per capita throughout the period under examination, avoiding the declining trends of EI identified for China after the 1970s, and for South Korea in the first decade of the 21st century.

The eight countries with short-run data (India, Mexico, Argentina, Turkey, Pakistan, Malaysia, Canada, and Australia) present different trends for E/GDP and E/Utility. The EI of the economies of India, Mexico, Argentina, Turkey, Pakistan, and Malaysia follow increasing trends according to the E/Utility indicator throughout the period with data available. In Canada, E/Utility increased until the 1970s and then decreased until 2000 when its EI stabilized. For Australia, the E/Utility had been increasing until the mid-1990s when it stabilized, while after 2005 it was declining. On the contrary, the E/GDP ratio indicates trends of relative stability (Brazil, Mexico, Turkey, Pakistan and Malaysia), decreases (Canada and Australia) or mixed trends with signs of decrease in recent years (India and Argentina). Notably, in all economies with different trajectories between E/GDP and E/Utility, the "per capita energy" use follows trends that share substantial similarities with the E/Utility ratio. An exceptional case in this respect is India, where per capita energy use increased intensely after 2000, although its E/Utility had stabilized in the same period.

The trends in per capita energy use demonstrate an essential aspect of the link between energy and the economy, since per capita energy is consumed through the use of goods and services produced by the economic process. In all economies studied, per capita energy use demonstrates closer similarities with the trends of E/Utility than with the trends in E/GDP. Within this context, detailed estimates for the three economies with long run data indicate a drastically increasing difference between per capita energy use and E/GDP in comparison to the difference between per capita energy and E/Utility (Figs. 6.1, 6.3 and 6.5). Furthermore, the economies of Japan, India, China, Indonesia and South Korea are interesting cases because their per capita energy use increases rapidly until the end of the period with data available, a trend that does not follow the stability or declining paths of E/Utility. Nevertheless, even in these cases, the trajectories of per capita energy better approximate those of E/Utility in comparison to those of the E/GDP index. These findings are important. They suggest that, although the increasing efficiency of the economic process induces a delink between the economy and energy, the increasing utility enjoyed by citizens requires substantial energy inputs which make the economic process dependent on energy flows. Increasing utility is depicted in the individuals' consumption of a basket of goods with relatively higher monetary value. The value of the basket increases and its composition shifts towards service-like and knowledge-based goods. The energy required for the production of goods making up the "representative basket" and consumed by the average citizen is increasing; this is reflected in the trends of the per capita use of energy.[1] Within this intricate context, the ultimate outcome of the economic system requires substantial energy inputs, which cannot be approximated by the amount of energy required for the one unit of monetary value reflected by the E/GDP ratio. The amount of energy required by one unit of utility, which shares more similarities with the amount of energy consumed by one citizen in the enjoyment of utility than the amount of energy required by one unit of monetary value, stands as a satisfactory approximation to EI. These similarities offer a validation of the suitability of E/Utility as an index that evaluates the actual link between energy (the fuel of the engine) and the economy (the engine which produces utility).

---

[1]Note that in our estimates of the per capita energy use, we estimate only the energy that is consumed domestically by the average citizen. These estimates include the energy directly consumed by the average citizen through the basket of goods he "consumes" (hence, the services these goods provide), and the energy use which is indirectly embodied in those goods that are produced domestically. However, our estimates do not include the energy embodied in imported goods. The embodied energy of imported goods (hence, the energy consumed during their production process) is an issue of paramount importance which, however, remains beyond the estimates and the scope of the present volume.

7.2 Coupling and Decoupling Economies

In this section, the economies under examination are classified according to the trends of the E/Utility index. The EI of the economies is examined over three time periods which were defined by two milestones for energy use: the oil crisis of the mid 1970s, which induced initiatives for advancing energy efficiency, and 2000 which initiated substantial restructuring of the economies towards knowledge-based sectors (financial services, social media, telecommunications, etc.). Indeed the empirical estimates confirm these milestones.

Until the mid-1970s, all of these economies follow an increasing trajectory for the rate E/Utility. It is not surprising that the intensities differ between countries. The heavily increasing energy intensity of the global economy and Japan until the first oil crisis is noteworthy.

The period between the mid-1970s and 2000 is characterized by different paths for the EI of the economies studied. In the 1970s a decoupling is initiated for the economies of the Netherlands, Norway, Italy, the USA, China, Japan, Germany, and the UK. This decoupling prevailed until 2000, with the exception of Japan where EI was stable after the end of 1980s. Australia indicates a relative stability for the period between the mid-1970s and 2000, while Canada followed an increasing trend until 1996 when it initiated a substantial decoupling. All the remaining economies continued the increasing trends of their EI until 2000, trends which had prevailed prior to the 1970s. Notably, all the economies with an intensified EI after the first oil crisis and until 2000, belong to the so-called developing countries, with the exception of Canada which, however, demonstrates a decoupling after the mid-1990s. This classification is based on the GDP per Capita index which ranges between 640 (India in 1970) and 8100 (Argentina in 1975) US$-1990GK.[2] The global economy reaches a GDP per Capita of approximately 4,000 US$-1990GK in 1975 and continues to increase its EI for the whole period between the first oil crisis and 1997. (Annex III presents the indexed trends of GDP and GDP per Capita for all the economies studied). On the other hand, all the economies that initiate decoupling or display a stable relationship between growth and energy around the middle of the 1970s have a per Capita GDP above 11,000 US$-1990GK and all belong to the so-called developed countries. The same holds true for Germany and the UK which initiated the decoupling trend earlier, around the beginning of 1970s, with a GDP per Capita over 10,000 US$-1990GK in the very same period. Strangely enough, China penetrates the "decoupling group", as it initiated a declining EI around 1977, with a per Capita GDP at the level of approximately 691 US$-1990GK. China possesses an additional particular characteristic; its per capita energy continued to increase after 1977 in a direction opposite to that of its EI. Overall, the case of China presents peculiarities which make it a distinct case. The trends of the EI in China call for a close analysis which takes into account the

---

[2]We indicate the minimum and the maximum values of GDP per Capita for the whole period, until 2000.

particular structure of China's economy and society. Probably international trade and distributional issues could offer some sources of explanation.

The period after 2000 is characterized by prevailing decoupling trends for the majority of the economies studied. Notable exceptions are Brazil, Mexico and Thailand which continue to increase their EI during 2000–2013. The EIs of Turkey and Malaysia had also been increasing since 2000; however, by 2010, they appeared to stabilize. Developing India displays a constantly stable EI from 2000 until the end of the period examined. All countries with increasing and stable EIs also present growing per capita energy consumption after 2000. Remarkably, all these countries belong to the group of developing economies with a GDP per Capita ranging from 1,900 US$-1990GK for India (in 2000) up to 7,700 US$-1990GK for Malaysia (in 2000).

The first decade of the 21st century constitutes a milestone for the EI of the global economy as well. The EI of the global economy peaks in 1998 and then stabilizes until the end of the period with data available; nevertheless per capita energy use increases throughout this period. The global GDP per Capita reaches 6,018 US$-1990GK, in 2000 when, indicatively, those of the USA and Japan are at the levels of 28,800 and 20,500 US$-1990GK, respectively.

The empirical estimates suggest a concrete relationship between EI and the level of utility, as measured by the GDP per Capita index. The EI of an economy probably tends to decrease once the GDP per Capita is above a certain level. Indeed, the analysis indicates that 7,000 US$-1990GK is the minimum necessary level for an economy to initiate a declining shift in EI. With the exception of the giant economy of China, all other national economies initiate a decoupling trend only when the GDP per Capita surpasses the level of 7,000 US$-1990GK. This by no mean implies that each and every national economy is able to decrease its EI once it reaches this level. Put simply, the empirical data suggest that with the exception of China, no other economy initiated a decreasing EI before reaching a GDP per Capita of 7,000 US$-1990GK. China, as already noted, constitutes an economy with very specific characteristics regarding the link between growth and energy use. Indeed, China's economy calls for an in-depth analysis which falls beyond the scope of the present volume.

7.3   Two Interesting Cases of Decoupling

The economies of Germany and the UK indicate a decoupling trajectory for EI, almost from the beginning of 1970s until the end of the period with data available (2013). As indicated in Table 7.1, one unit of utility was produced with substantially reduced energy inputs between 1970 and 2013. Specifically, the E/Utility rate declined from 30.38 (1970) to 15.03 (2013) toe/Utility$ in Germany and from 20.12 (1970) to 8.24 (2013) toe/Utility$ for the UK. Notably, in 2013, the energy requirements of one unit of utility are far lower in the UK and Germany than in the advanced economies of the USA (70.29 toe/Utility$ at 2013) and Japan

Table 7.1 Absolute values of a set of energy relevant indicators for the UK, Germany, the USA, and Japan

	1970	1980	2000	2013
(a) *DEC/GDP* (toe/$/yr)				
Germany	0.39	0.34	0.22	0.18
UK	0.36	0.28	0.18	0.13
USA	0.53	0.43	0.28	0.22
Japan	0.28	0.23	0.20	0.16
(b) *DEC/Utility* (toe/Utility$/yr)				
Germany	30.38	26.91	17.92	15.03
UK	20.12	15.58	10.39	8.24
USA	108.30	97.57	80.25	70.29
Japan	28.82	26.48	25.29	20.87
(c) *DEC per capita* (toe/person/yr)				
Germany	3.98	4.54	4.04	3.99
UK	3.89	3.58	3.78	3.15
USA	7.94	7.96	8.18	7.14
Japan	2.68	3.04	4.09	3.72
(d) *DEC* (mtoe/yr)				
Germany	309.6	355.9	333	325
UK	216.7	201.4	224	200
USA	1,627.7	1,812.6	2,313.7	2,265.8
Japan	279.9	355.6	518	474
(e) *GDP per capita* ($/person/yr)				
Germany	$10,190.09	$13,222.25	$18,581.44	$21,624.43
UK	$10,767.47	$12,931.49	$21,547.76	$24,267.31
USA	$15,029.85	$18,577.37	$28,830.57	$32,235.88
Japan	$9,713.95	$13,427.73	$20,477.06	$22,708.89

(20.87 toe/Utility$ at 2013). The standard E/GDP indicator shares similar decoupling trends with the E/Utility ratio. In addition, per capita energy use (see Table 7.1) further supports the decoupling trends indicating, between 1970 and 2013, stability in the case of Germany and a decline in the UK.

These entire developments led to a substantial decrease in total energy use in both economies. The total energy peak occurred at the end of the 1970s and the beginning of the 2000s for Germany and the UK, respectively, and subsequently a decrease was noted until 2013. Evidently, total energy consumption is influenced by population size. Population size decreases smoothly after 2003 in Germany whereas it increased at a slow pace throughout the period examined in the UK. Simultaneously, both economies substantially increased their ultimate outcome, utility, with per Capita GDP increasing by 125.4 % for the UK, and 112.2 % in Germany, between 1970 and 2013 (Table 7.1).

All these findings suggest that the economic systems of Germany and the UK have experienced good performance both in economic and energy terms. They have provided increasing economic utility with constantly decreasing energy intensity. The trajectories of all the energy relevant indicators suggest two economic systems which are gradually gaining relative "independence" from energy flows.

Several factors drive the economic and the energy performance of these economies. The effects of technological efficiency, the shift towards service-like and knowledge-based goods, and the outsourcing of heavy industries strongly influence their energy profile. The evaluation of the exact weight of each factor requires an analytical endeavour which is beyond the scope of the present volume. Nevertheless one can firmly assert that the economies of the UK and Germany constitute two distinct cases which may offer good lessons at least for the so-called developed countries. Furthermore, these two economies constitute two important analytical 'experiments' for the essential evaluation of the energy-economy link. The relevant empirical estimates demonstrate that the Energy Intensity of an economic system is a complicated issue whose evaluation requires the investigation of several factors, depicted through a number of indicators. The evaluation of energy intensity solely on the basis of E/GDP and E/Utility ratios may be a shallow approach which leads to misleading conclusions. In this context, we may safely assert that trends in "Energy per capita" and "total energy use" offer a better insight into the essence of the actual link between energy and the economy.

7.4 Re-evaluating the Energy Intensity of the Economy. Concluding Remarks

Energy is an indispensable input of the production process and hence of the economy. The historical evolution of the production process prescribed different patterns for the energy-economy link in different periods of economic history (Smil 1994; McNeill 2000; Heinberg 2011). This link attracted the interest of economic analysis since it is related to the pressing problem of the dependency of the economy on natural resources, an issue that implicitly or explicitly penetrates contemporary economic thought. The "optimistic" school asserts that the power of technological progress will free the economic process from its dependency on natural resources and hence on energy. Technological advances make it possible to substitute other factors for natural resources (Solow 1974). On the contrary, the "pessimistic" school argues that the entropic nature of the economic process establishes an irrevocable dependency on natural resources (Georgescu-Roegen 1971, 1975, 1982; Daly 1997). Only recently has the availability of reliable long run datasets permitted an in-depth empirical investigation of the link between the economy and resources. The first findings seem to support the arguments of the "optimistic" school. Indeed, the prevailing index of E/GDP, which has been used for the evaluation of the Energy-Economy link, has been declining for the majority of the economies studied throughout the economic history of 20th century and until recently (2013). This is the so-called decoupling effect supporting a momentous decrease in the dependency of growth on energy resources.

The present volume questions these findings by proposing an alternative core indicator for the link between the economy and energy. We argue that the actual energy requirements of the economy can only be approximated through the energy inputs required for the creation of one unit of utility. Utility is the ultimate outcome of the economic system. And the energy intensity of the economy can be evaluated by comparing energy inputs with the actual outcome of the economy. Hence, energy intensity is evaluated at the border of the economic system: comparing energy inputs with utility outputs.

Utility is an individualistic perception which is created through the consumption of goods produced by the economic process. The creation of utility requires actual goods which satisfy human needs. Human needs endow economic goods with certain physical properties. These properties make it necessary to use substantial energy flows for the production of real world goods. Energy efficiency can reduce the energy requirements but this trend is limited both by the physical properties of goods and the technical efficiency of energy use. In this context we investigate the energy requirements for the creation of one unit of GDP per Capita, since this is the only available monetary-based index that has been broadly used for the evaluation of economic utility. Indeed, GDP per Capita approximates the utility enjoyed by human beings and is therefore used for ranking and classifying economies on the basis of their performance. We estimate the energy intensity of 22 representative economies which constituted 76.9 % of the global economy in 2013 in terms of GDP, expressed in 1990 million GK US\$ (The Conference Board Total Economy Database 2014), as well as the global economy, through the **E/Utility = E/[GDP per Capita]**, and compare these estimates with those of the standard E/GDP ratio. The findings demonstrate that, at the global level over the last one hundred years or so, economic growth was critically dependent on the consumption of energy. The unprecedented increasing economic utility of the 20th century was an energy squanderer since an additional unit of GDP per Capita required a disproportionate increase in energy use. Only recently, since 2000, does the energy intensity of the global economy indicate a pause in these increasing trends. Similarly, the two technologically advanced economies (the USA and Japan), with data available for the Economy-Energy link for the entire 20th century, follow an increasing dependency of growth on energy until the mid-1970s when the oil crisis induced energy saving initiatives which resulted in a substantially decreasing EI. Notably, in very recent years, after 2000 or so, the energy intensity of these knowledge based economies indicates suggestions of relative stability that appear to interrupt the decreasing trends. These findings, taken in conjunction with the relative stability of the EI in a number of developing countries in recent years, as well as with the increasing EI in another set of large economies (Brazil, Mexico, India, etc.), cannot fuel optimism over the independence of the contemporary economy from energy flows. Nevertheless, a small number of developed countries have consistently displayed a decoupling of growth from energy. These countries offer positive examples and suggest that potentially the energy requirements of a knowledge-based economy can be substantially reduced. However, this evolution of a number of knowledge-based economies does not constitute a real breakthrough for the Economy-Energy link at the global level. As the decoupling trends of certain knowledge-based economies could be partially attributed to international trade, recent trends of EI at the global level, as well as in the developing world, could be seen as a counterbalance to the decreasing EI of these knowledge-based economies.

Overall, recent trends suggest a shift in the link between the economy and energy. However, this shift does not support a fundamental de-linking. This conclusion gains support from other directions of research on the Energy-Economy link. Ayres and Warr (2009), Warr et al. (2010), and Ayres et al. (2013) argue, on the basis of the systematic empirical analysis of a number of national economies, that energy is the irreducible engine of growth. Similarly, recent reviews of the energy-growth causality suggest that, although the direction of causality is case-dependent, there exists an irrevocable link between energy and growth (Kalimeris et al. 2014).

In recent years, mankind has been facing the transition from a fossil fuel based economy to something else which is still not clearly defined. Certain scholars (Rifkin 2011) assume that internet technology and renewable energy are merging into the "Third Industrial Revolution", with an economy free from constraints imposed by energy scarcity. However, the replacement of fossil fuels is still only a wish and an expectation. It can only be speculated what the world beyond fossil fuels would look like. The question then remains: Is mankind's economy approaching an era that will be independent of energy resources, induced by a gradual but essential decoupling of the economy from energy flows? If production is perceived as a dimensionless abstraction measured in monetary units reflected by aggregate GDP, then there are high expectations for such independence. On the other hand, if economic production is envisaged as a mixture of goods and services facilitating human needs and therefore having certain physical properties then, although there may be some potential for decoupling, irrevocable constraints also exist because of the physical properties of real world products. This places the academic dialogue between growth optimists, a-growth, and de-growth supporters, within a more realistic framework (van den Bergh 2011; Kallis 2011; Kallis et al. 2012; Victor 2012).

References

Ayres, R. U., van den Bergh, J. C., Lindenberger, D., & Warr, B. (2013). The underestimated contribution of energy to economic growth. *Structural Change and Economic Dynamics, 27,* 79–88.

Ayres, R. U., & Warr, B. (2009). *The economic growth engine: How energy and work drive material prosperity.* UK: Edward Elgar Publishing.

Daly, H. E. (1997). Georgescu-Roegen versus Solow/Stiglitz. *Ecological Economics, 22*(3), 261–266.

Georgescu-Roegen, N. (1971). *The entropy law and the economic process*. Cambridge: Mass.

Georgescu-Roegen, N. (1975). Energy and economic myths. *Southern Economic Journal, 41*(3), 347–381.

Georgescu-Roegen, N. (1982). Energetic dogma, energetic economics, and viable technologies. *Advances in the Economics of Energy and Resources, 4*, 1–39.

Heinberg, R. (2011). *The end of growth: Adapting to our new economic reality*. New Society Publishers.

Kalimeris, P., Richardson, C., & Bithas, K. (2014). A meta-analysis investigation of the direction of the energy-GDP causal relationship: implications for the growth-degrowth dialogue. *Journal of Cleaner Production, 67*, 1–13.

Kallis, G. (2011). In defence of degrowth. *Ecological Economics, 70*, 873–880.

Kallis, G., Kerschner, C., & Martinez-Alier, J. (2012). The economics of degrowth. *Ecological Economics, 84*, 172–180.

McNeill, J. R. (2000). *Something new under the sun: An environmental history of the twentieth-century world* (the global century series). New York: WW Norton & Company.

Rifkin, J. (2011). *The third industrial revolution: How lateral power is transforming energy, the economy, and the world*. New York: Palgrave Macmillan.

Smil, V. (1994). *Energy in world history*. Boulder: Westview Press.

Solow, R. M. (1974). The economics of resources or the resources of economics. *The American Economic Review, 64*(2), 1–14.

The Conference Board Total Economy Database. (2014). Available at: http://www.conference-board.org/data/economydatabase/.

van den Bergh, J. C. J. M. (2011). Environment versus growth—A criticism of "degrowth" and a plea for "a-growth". *Ecological Economics, 70*(5), 881–890.

Victor, P. A. (2012). Growth, degrowth and climate change: A scenario analysis. *Ecological Economics, 84*, 206–212.

Warr, B., Ayres, R., Eisenmenger, N., Krausmann, F., & Schandl, H. (2010). Energy use and economic development: A comparative analysis of useful work supply in Austria, Japan, the United Kingdom and the US during 100 years of economic growth. *Ecological Economics, 69*(10), 1904–1917.

Annexes

Annex I: Percentage Change of the EI Indicators and Per Capita Energy Use

Table A.1 Percentage change of DEC/GDP ratio for indicative periods

Countries	1965–2013 (%)	1965–1980 (%)	1980–2007 (%)	2008–2013 (%)
Developed				
Australia	−37.04	10.55	−27.74	−20.06
Canada	−32.41	−0.94	−28.29	−3.78
Germany	−53.28	−13.03	−44.13	−3.49
Italy	−28.38	−3.85	−21.12	−5.75
The Netherlands	−27.46	18.85	−33.93	−3.98
Norway	−38.89	−7.19	−28.50	−9.74
South Korea	51.76	58.37	−2.62	−1.27
Spain	−4.99	17.61	−10.62	−6.73
Taiwan	−28.54	10.63	−22.09	−11.75
United Kingdom	−65.04	−25.51	−49.72	−8.00
Developing				
Argentina	−5.08	−11.18	12.60	−4.03
Brazil	84.63	31.94	31.23	6.63
China	−33.25	52.77	−51.81	−9.34
India	−9.75	14.04	−17.73	−3.41
Indonesia	82.71	36.93	45.30	2.03
Malaysia	78.63	35.92	48.38	−7.60
Mexico	35.14	21.02	12.18	−0.68
Pakistan	−12.85	−11.78	12.16	−7.06
Thailand	174.06	65.13	53.76	6.78
Turkey	40.88	16.09	24.18	−1.23

© The Author(s) 2016
K. Bithas and P. Kalimeris, *Revisiting the Energy-Development Link*,
SpringerBriefs in Economics, DOI 10.1007/978-3-319-20732-2

Table A.2 Percentage change of DEC/Utility ratio for indicative periods

Countries	1965–2013 (%)	1965–1980 (%)	1980–2007 (%)	2008–2013 (%)
Developed				
Australia	22.52	41.25	2.58	−15.28
Canada	16.42	21.38	−3.96	0.14
Germany	−49.72	−9.96	−41.14	−4.57
Italy	−15.26	4.41	−16.64	−3.57
The Netherlands	−0.92	36.76	−23.69	−1.67
Norway	−22.48	1.85	−19.01	−8.22
South Korea	158.82	110.34	23.25	−0.09
Spain	40.28	37.42	7.79	−3.77
Taiwan	29.75	54.16	−0.18	−10.55
United Kingdom	−59.16	−22.82	−45.24	−5.38
Developing				
Argentina	81.52	13.08	58.95	1.01
Brazil	351.91	95.34	105.25	11.84
China	26.69	109.60	−35.27	−2.87
India	128.65	59.66	36.89	3.39
Indonesia	327.22	90.68	128.38	7.70
Malaysia	437.06	93.91	183.85	0.07
Mexico	247.93	83.23	78.41	4.98
Pakistan	182.53	30.76	122.79	0.63
Thailand	476.53	142.20	112.89	9.90
Turkey	255.81	63.68	106.11	5.15

Table A.3 Percentage change of DEC per capita ratio for indicative periods

Countries	1965–2013 (%)	1965–1980 (%)	1980–2007 (%)	2008–2013 (%)
Developed				
Australia	72.17	56.94	27.65	−13.18
Canada	66.22	53.01	12.16	−1.91
Germany	17.91	34.22	−13.44	0.57
Italy	67.70	66.40	19.10	−13.88
The Netherlands	76.13	78.37	11.24	−9.04
Norway	108.13	61.02	35.22	−6.05
South Korea	2,385.83	353.63	375.20	13.41
Spain	219.49	127.27	73.57	−15.89
Taiwan	883.05	221.44	218.02	1.96
United Kingdom	−13.00	−1.22	−0.51	−11.51
Developing				
Argentina	63.34	14.41	34.26	4.90
Brazil	425.51	179.95	57.78	15.02
China	1,043.90	131.04	236.15	41.20
India	345.36	38.85	146.30	24.49
Indonesia	914.33	160.59	222.95	27.84
Malaysia	1,007.85	175.50	289.06	4.63
Mexico	198.61	106.61	41.50	1.84
Pakistan	196.29	32.87	130.16	−0.96
Thailand	2,019.48	222.49	429.42	21.57
Turkey	416.35	86.45	156.37	9.95

Annex II

See Fig. A.1.

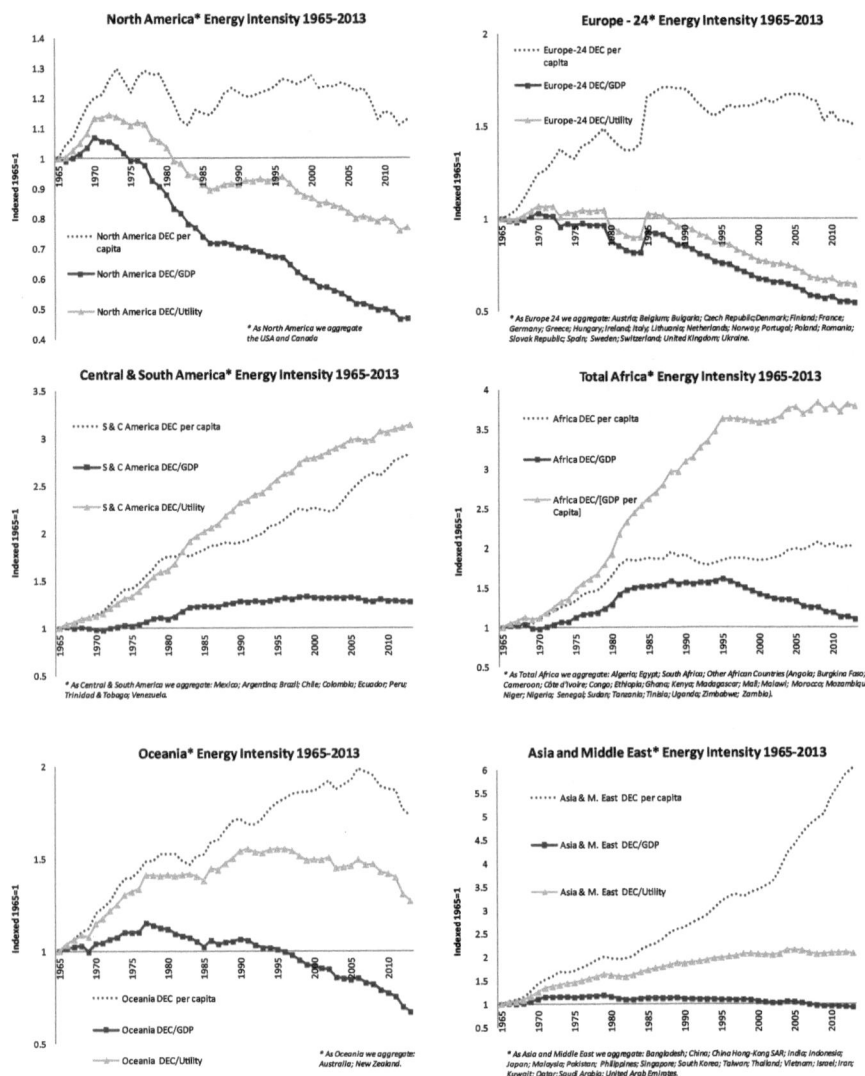

Fig. A.1 Energy intensity at the continental level

Annex III: Socio-economic Indicators and Energy Inputs

See Figs. A.2, A.3 and A.4.

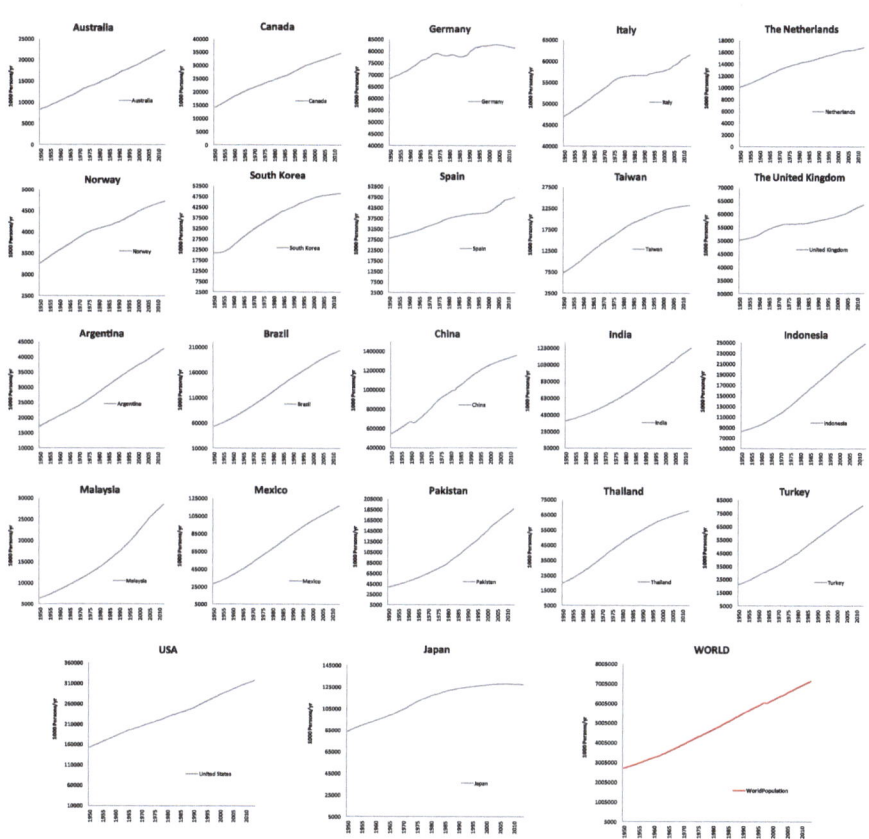

Fig. A.2 Population 1950–2013 (in 1000 persons/year)

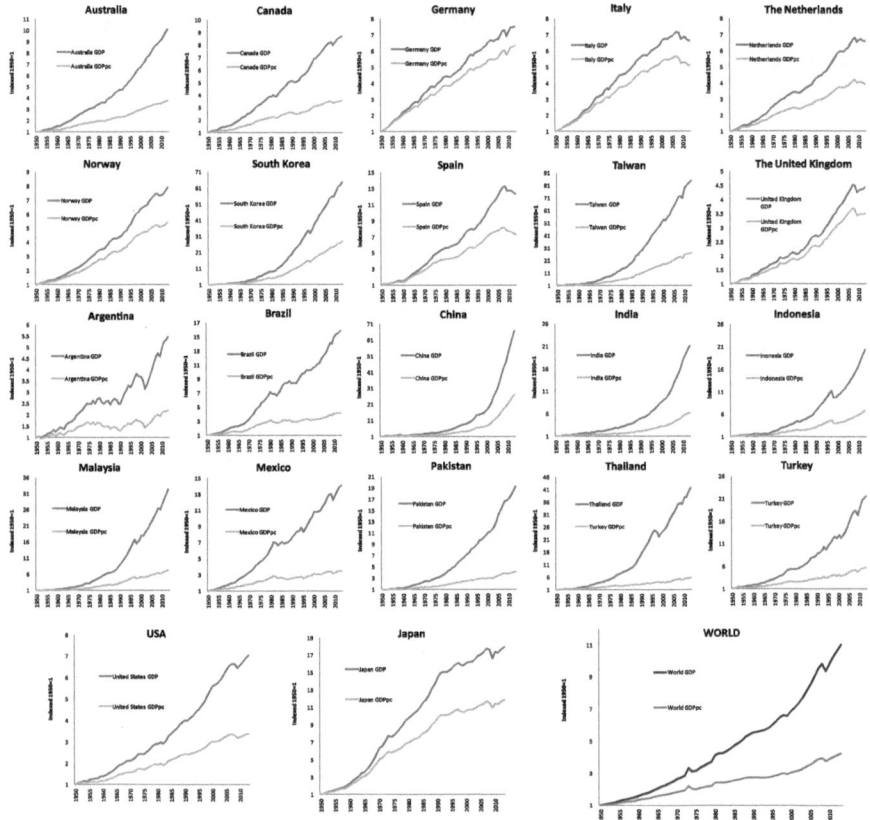

Fig. A.3 GDP and GDP per capita, for 1950–2013 (Indexed 1950=1)

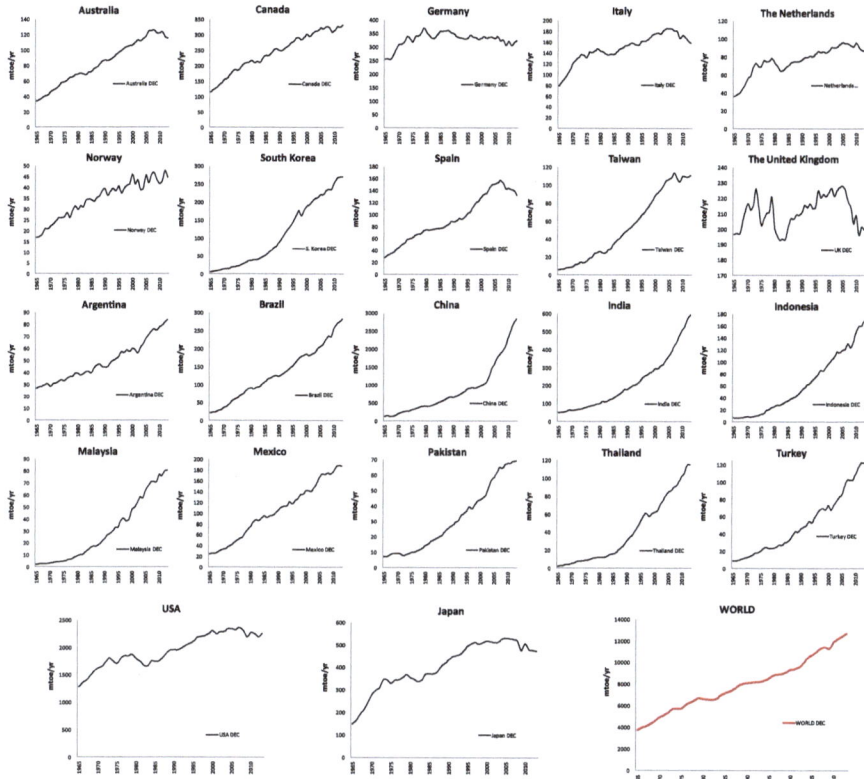

Fig. A.4 Total DEC for 1965–2013 (in mtoe/year)